Sustainable Development Goals Series

The **Sustainable Development Goals Series** is Springer Nature's inaugural cross-imprint book series that addresses and supports the United Nations' seventeen Sustainable Development Goals. The series fosters comprehensive research focused on these global targets and endeavours to address some of society's greatest grand challenges. The SDGs are inherently multidisciplinary, and they bring people working across different fields together and working towards a common goal. In this spirit, the Sustainable Development Goals series is the first at Springer Nature to publish books under both the Springer and Palgrave Macmillan imprints, bringing the strengths of our imprints together.

The Sustainable Development Goals Series is organized into eighteen subseries: one subseries based around each of the seventeen respective Sustainable Development Goals, and an eighteenth subseries, "Connecting the Goals", which serves as a home for volumes addressing multiple goals or studying the SDGs as a whole. Each subseries is guided by an expert Subseries Advisor with years or decades of experience studying and addressing core components of their respective Goal.

The SDG Series has a remit as broad as the SDGs themselves, and contributions are welcome from scientists, academics, policymakers, and researchers working in fields related to any of the seventeen goals. If you are interested in contributing a monograph or curated volume to the series, please contact the Publishers: Zachary Romano [Springer; zachary.romano@springer.com] and Rachael Ballard [Palgrave Macmillan; rachael.ballard@palgrave.com].

Rasmus Rodineliussen

Underwater Worlds

An Ethnography of Waste, Pollution, and Marine Life

palgrave
macmillan

Rasmus Rodineliussen
Department of Social Anthropology
Stockholm University
Stockholm, Sweden

ISSN 2523-3084 ISSN 2523-3092 (electronic)
Sustainable Development Goals Series
ISBN 978-3-031-63369-0 ISBN 978-3-031-63370-6 (eBook)
https://doi.org/10.1007/978-3-031-63370-6

This Palgrave Macmillan imprint is published by the registered company Springer Nature Switzerland AG.
The registered company address is: Gewerbestrasse 11, 6330 Cham, Switzerland

Photo: Oskar Kihlborg

To Lise, Lili, and Leo

Acknowledgments

Writing this book has been an extended learning experience. Many hours lie behind these pages—my own and others—and I care deeply about the topics treated here. The list of people who have made this project possible is long, and although I cannot mention all of you here, I hope you know how much your support and assistance have meant to me.

There are, however, a few individuals that I feel bound to acknowledge particularly. Fredrik Johansson, you are the one who introduced me to the world of trash scuba diving in Sweden, and who has become a close friend over the years. Thank you for everything! I hope we will have many memories yet to share with the amazing group of individuals who make up Rena Mälaren: a special mention of Markus Olsson for your care of my writing. Paula Dalcin Martins, Andy Booth, Lisbeth Sørensen, and Rasa Slizyte, you have all invested a great deal of energy in my work and have generously provided me with insights into your laboratory worlds. And you, Martijn Hermans, have given me much time, answered many questions, and have become a close friend. Together with the other scientists, I thank you all sincerely for sharing your knowledge and experience with me, and thereby making this research possible. I also must acknowledge the kindness of the Baltic Sea Center at Stockholm University, which allowed me on board one of their research vessels. There I met the most amazing captain, Thomas Strömsnäs, who made my time at sea such an eye-opener. Thomas, you hoped I would write something about *Electra* in the end—so here you are!

I have had the great honor of being supervised by Bengt G. Karlsson and Helena Wulff, two eminent scholars, colleagues, and friends. I know working with me has not always been easy, receiving draft after draft from me (too many at times, I know—sorry!). And still, you have both remained supportive, critical, and enthusiastic. I could not have asked for better supervision, and I also have found myself two close friends for the future. Another colleague and friend deserving special mention is Paula Uimonen, who has given me a great deal of assistance. It is incredible to have such people around you!

I cannot say enough for the Department of Social Anthropology at Stockholm University. It has been an inspiring space in which to develop as an academic. I have had many wonderful colleagues there who took part in seminars and provided a superb intellectual environment: Ulf Hannerz, Annika Rabo, Shahram Khosravi, Renita Thedvall, Johan Lindquist, Eva-Maria Hardtmann, Karin Ahlberg, Tomas Cole, Simon Johansson, Camelia Dewan, Hege Høyer Leivestad, Molly Sundberg, Anna Gustafsson, Raoul Galli, Gudrun Dahl, Igor Petričević, Ulrik Jennische, Andrew Mitchell, Ivana Maček, Anna-Karin Olsson, Isabella Strömberg, Jonathan Krämer, Victor Nygren, and Elin Linder. Elin has been especially helpful in commenting on several early drafts of this manuscript. The department has also organized a midway and final seminar for my dissertation that became this book, with Mark Graham (midway) and Thomas Hylland Eriksen (final) as opponents. Both have provided close readings of my work and assisted me greatly in improving my writing and thinking.

Another colleague and friend whom I met for the first time in Seattle 2022, at the American Anthropological Association's annual meeting, although we had been in almost daily contact since the fall of 2019, is my co-editor at the *Anthropology Book Forum*, Emilia Groupp. Your thoughts and our conversations have inspired my thinking considerably. Moreover, I wish to thank Teddy Primack for making the sometimes tiresome work of language editing into a fun and collaborative experience. You really did more for my writing than your service required, also including the assistance of your partner, Anna Lindberg. It has been a pleasure to know and work with you.

I also want to thank the editors at Palgrave Macmillan, Elizabeth Graber and Sarah Hills, who both have been accommodating and supportive. It has been a great experience to work with you both. Moreover, I want to thank the reviewers of this book for their clear and guiding comments. As this book builds on my dissertation I further want to thank my opponent Marianne Elisabeth Lien, and the examining committee—Jon Henrik

Ziegler Remme, Maris Gillette, Andrew Johnson, and Lowe Börjeson—for their critical comments.

This book was made possible through generous grants for fieldwork and conference attendance from the Swedish Society for Anthropology and Geography (SSAG), the Forum for Asia Studies, the Helge Ax:son Johnson Foundation, and the Stockholm University Donationsstipendier funding, which have been of crucial importance. I also want to thank the Swedish tax payers who supported me through my salary for close to five years of doctoral studies.

I am very fortunate to be surrounded by a wonderful family and friends. My mother Lise has provided boundless support and has always remained curious about what it is I do. I must not forget to mention my two small helpers, Lili and Leo, who have accompanied me to Brazil and Thailand, as well as having pulled and carried dirty objects from Swedish waters. My children have framed my work in many ways. On a picture I have on my wall at home Lili has painted a diver and written below: "My dad saves the world." Even if I am not convinced that I am saving the world, it means a lot to me that my children think so.

Nacka, Sweden
11 March 2024

Rasmus Rodineliussen

Competing Interests There are no competing interests.

Ethics Approval All interlocutors included in this book have been informed about the purpose of the study and future publications, and have given their informed consent to participation prior to their involvement.

Praise for *Underwater Worlds*

"*Underwater Worlds* takes readers on a deep dive into the entanglements of care, technology, and contamination in contemporary water-worlds. Richly granular narratives and innovative more-than-human methodologies undergird the work's innovative articulation of "aquabiopolitics" as a conceptual lens into the relationship of waste, pollution, and marine life, enriching and informing fields including environmental anthropology, political ecology, STS, and the environmental humanities."

—Sophie Chao, Lecturer in Anthropology, *University of Sydney*

"With great sensitivity and scholarship, Rodineliussen shows us how, more than a sink or a bath, our waters have long been treated as a toilet in which we dump our shit, from industrial effluents, to plastics and old car batteries. This (sea) bottom-up ethnography of renegade citizen-scientists offers a compelling vision of a new water politics that moves us from bare life to thriving ecosystems."

—Patrick O'Hare, Senior Researcher, Department of Social Anthropology, *University of St Andrews*

"Underwater ethnography will be adopted widely in cultural anthropology, cultural ecology and ethnographic methods courses, plus courses in underwater archaeology and material culture. Rodineliussen's vivid, accessible description of scientists' and divers' worlds and where they intersect is critical in generating new approaches to environmental challenges. His emphasis on scientific collaboration is key, also making his approach valuable to scholars and students in environmental studies and marine biology. This book is inspiring for its descriptions of the creative and dramatic ways Rodineliussen's interlocutors address the public with calls to action, and for how his own work can help point the way for activists in a wide range of settings."

—Maria D. Vesperi, Professor of Anthropology, *New College of Florida*

Contents

About the Author

Rasmus Rodineliussen holds a PhD from the Department of Social Anthropology at Stockholm University. His work focuses on the relationship between humans, waste, pollution, and water—the topic of this book. Earlier, Rasmus worked with social online activism and migration, mainly concerning the war in Syria. Rasmus has published in a number of peer-reviewed journals, such as *Anthropology Now, Anthrovision*, *Visual Studies*, the *Irish Journal of Anthropology*, and *Kritisk etnografi*. He has also written two method case studies for *SAGE*. Rasmus has been awarded the General Anthropology Division of the American Anthropological Association's "New Directions Prize" for 2022 in the group category for his work as co-editor for the *Anthropology Book Forum*.

List of Figures

CHAPTER 1

Diving In

> Now this was something new! This is by far the worst visibility I have ever encountered. There is no visibility, none at all! I consider myself a diver with some experience, so if *I think this was scary*, then this is going to be interesting today. Not sure that everyone in the group will follow through with this dive.

These are the words of Fredrik Johansson, a military-trained diver who has spent his adult life diving within the notoriously bad visibility of Lake Mälaren in Stockholm, Sweden. The other divers he refers to are part of the non-profit organization of trash scuba divers that he has founded in Stockholm that go by the name *Rena Mälaren*. I got to know him during fieldwork in and around Stockholm for this **book** between 2019 and 2024. On the day Fredrik referred to above, the organization held a visit-cleanup dive in Västerås, a small city outside Stockholm. Let me take you there....

The water below us, a steady 5 degrees Celsius, did not look like ordinary water; it had a brownish-orange color. One diver next to me said that it looked like a cafe latte. When we eventually submerged underwater, we faced a brownish-orange wall in front of our goggles. The line between air and water, over the surface or under, was clearly defined. Two different worlds! Once submerged, all one could see was what was within the scuba goggles, as visibility ended right with the glass. We knew beforehand that

R. Rodineliussen, *Underwater Worlds*, Sustainable Development Goals Series, https://doi.org/10.1007/978-3-031-63370-6_1

the water was shallow and would not be much deeper than 3 meters, in fact I quickly touched the bottom with my fins. Then I slowly lowered my body so that I was lying flat right above the bottom. This had to be done in a controlled manner as there could very well be sharp objects in the way. Once safely on the bottom, it took me a moment to mentally and physically adapt to the situation. Here I was, submerged in water, with absolutely no way of seeing what was in front, beside, or behind me. The risk of encountering sharp objects in the water was immense—we were, after all, there to look for just such things. I calmed my breathing down and engaged my other senses, such as touch, feeling with hands and fins, and listening, because sound underwater has a very particular way of vibrating through the body, coming from everywhere at once. For a diver, sound is very important, particularly when diving where one can encounter boats that can easily hit an undetected diver who is too close to the surface. Boat engines roar underwater, and this is a clear sound of danger. But contrary to one's first impulse to quickly move away—to escape—one ought to press oneself toward the bottom floor, to stay as deep down as possible until the roaring boat has disappeared. When you cannot see, your other senses become your connection to the world. Down here, in the cafe latte water, my only way of encountering the world was through these other bodily senses.

I started to move my left hand forward before sweeping it slowly to the side, reaching out for unknown objects in the water. Then I did the same thing with my right hand, then with my legs, one at a time. Only after not finding anything in any direction did I crawl forward a fraction before repeating the procedure. After a while, my left hand encountered a circular object. I touched it carefully before deciding that it felt like the wheel of a bicycle, and indeed there was a metallic frame! In my right hand I held a rope that connected me to the surface. I tied this rope to the bicycle—still without seeing anything—before following the rope upward to the surface. As I pierced the surface, murky dark turned into clear air and blue sky. It was very much like coming out of a pitch-black room. I signaled for those on land to pull, and up came the bicycle. The rope was thrown back, and I returned into the blinding confines of the water. I continued in this way, up and down, as I located new objects on the bottom. That day we were 11 divers crawling around on the bottom, looking much like blind frogs, with arms and legs moving in all directions. The divers moved like this in order to sensuously hunt for metal, car batteries, car tires, and other types of material waste in the water. The total catch of the day was about three tons of waste, stored in a container (Fig. 1.1).

In this **book**, I begin from the understanding that the underwater world is unknown to most people. This is because the human body is not made to be underwater for an extended period of time, but also because water and its surface can create a border between what is visible and what is not, especially in dark waters as in Sweden. This circumstance has created a convenient place to discard waste and pollutants, with damaging effects on marine life. Such is the situation in the Baltic Sea—perhaps more so than in other seas, and it has given the Baltic the reputation of being one of the most polluted seas in the world. The Baltic Sea, as one of the "most polluted" seas, stands in sharp contrast to the green image promoted of Sweden in various contexts, not least when it comes to water. As will be elaborated below, Lake Mälaren, the third-largest lake in Sweden, is in a similar distressed condition as the Baltic Sea. This contrast provoked my interest and motivated me to focus on the waters around Stockholm, Sweden's capital, where Lake Mälaren and the Baltic Sea meet, mixing fresh and brackish water. Stockholm is a special site with regard to water. The city has hosted the Stockholm Water Festival from 1991 to 1999 and handed out a water prize since 1991. Stockholm is also home to the Stockholm International Water Institute (SIWI), which annually sponsors one of the world's largest water conferences.

Fig. 1.1 (Left) Trash scuba diver in "cafe latte" water; (Middle) Electric scooter coming up from the water; (Right) The smelly and bulky "catch." (Photos: Author)

Making the Underwater World Visible: Aim and Research Questions

My objective is to open up the underwater world, and make its invisible waste and pollution visible.[1] I do this through *underwater ethnography*, where I join trash scuba divers on cleanup missions and publicity campaigns that combat the increasing pollution of the seas. I also follow marine scientists' as they seek to understand and combat the pollution. Through the concept of *aquabiopolitics*, I trace how marine life is made to serve human interests. Aquabiopolitics assists me in tracing interactions between humans, waste, pollutants, fish, microbes, and sea organisms. I have developed the term aquabiopolitics from Foucault's (1978) notion of "biopolitics," and the more recent work of scholars who are concerned with environmental issues, as will be discussed at length below.

The following chapters will explore ways of revealing what is beneath the surface of the sea in the underwater world. The story will focus on *how waste and pollution in the underwater world can be rendered visible, knowable, and relatable to humans.* Like many of the divers and scientists I follow, I am deeply concerned about the ongoing pollution and destruction of marine environments and my ultimate aim is to push for a larger societal debate about responsible ways to live with and care for water. I wish to

[1] The anthropology of "wastes" (Alexander and O'Hare 2023), show how human societies leak waste and pollutants in various shapes and forms into "nature." It may be from landfills or dumping (Reno 2015, 2016; Millar 2018; O'Hare 2022), animal and agricultural farms (Blanchette 2020; Rydén 2014; Waltz 2020), industry (Hoover 2017; Renfrew 2018; Dewan 2020), or other sources—making life today part of what Alex M. Nading (2020) has called "toxic worlding" (building on Chen 2012: 196). Leaking elements can either be small, almost invisible (nutrients or chemicals) or larger objects such as car batteries that leak pollutants such as lead. These larger objects can be understood by Michael Thompson's (2017) reasoning regarding "rubbish" as something between durable and waste, because of the capacity to return into the category of durables, although in another form, with human and other-than-human costs and gains (cf. Wenzel Geissler and Prince 2020; Minh and Nguyen 2018; Tousignant 2018). This leads to the realization that humans and other-than-humans now cohabit a toxic world. Our organisms are made by and with toxicants, such as POP's and microplastics (cf. Abasi et al. 2019; Ford 2019). Not all waste is seen as "bad" or "unwanted," but can instead be viewed and categorized as a commodity for industry to capitalize on (Alexander and O'Hare 2023). Hence, understanding the role of waste and pollutants in the world of today is complicated, as it can be rendered as toxic or a commodity, depending on your perspective and motif. Waste can, for example, be processed and used as fuel for cars as a complement to fossil fuels: are such waste good? Waste is tricky but also everywhere, and therefore, important to think with to understand the contemporary world.

understand how humans learn about and interact with the underwater world. With this engagement, this book asks: *(1) How can one show the human harm caused to the Baltic Sea and Lake Mälaren? (2) What is the role of technology in learning about the marine environment? And (3) in what ways can the practices of marine scientists and environmental activists impact multispecies care and prevent the destruction of aquatic worlds?*

While grappling with these questions, I consider the role of politics in and out of water, in *following the relations between humans, waste, pollution, and marine life*. These themes will run through the book and guide the argument. Theoretically I build on the anthropology of water, waste, political ecology, and Science and Technology Studies (STS), with inspiration from visual anthropology, and I include other-than-humans in my analysis. Methodologically, I use sensorial and embodied methods in a multi-sited field.

Situating the Study: Stockholm, the Baltic Sea, and Lake Mälaren

As someone born and raised in Stockholm, a city of about two million inhabitants located between the Baltic Sea and Lake Mälaren, I have always been proud to live in one of the few capital cities where the water is clean enough to drink from the tap, and where the inhabitants can even swim in the water around the city center. Stockholm is well known for this and boasts several beaches within walking distance from the middle of the city. Occasionally one reads in the local newspapers that this or that beach is temporarily unfit for swimming, but generally the water is assumed to be very clean.

As I was growing up, I began to take an interest in water and started to read about the history of Stockholm with relation to waste management and water. I found out that it was not until the 1900s that Stockholm's first wastewater treatment plant opened. Before that, all sewage was just dumped into the nearest body of water. In the 1700s, waste was poured out onto streets and alleys. Then by the 1800s, small ditches were constructed in some of the larger streets to lead the waste toward the water. There are stories about the stench that would ensue during hot summer days because of the polluted water and the dried-up filth in the ditches. Stockholm also has had a long history of being a city where factories dispensed all kinds of chemicals and metals into the water—much of which still lay buried in the sediment. There are places where you are not allowed to swim or even scuba dive unless you wear a chemical-protective dive suit together with a closed-circuit diving mask and regulator that will ensure you do not contact the water. This is within the areas where drinking water is taken: water treatment plants cannot clean all chemicals or metals from the water, which means that these pollutants eventually reach humans. The authorities do make tests, and the water is currently within the parameters to be classified as drinkable. However, during my fieldwork I have been told anonymously that the levels of pollution in Stockholm's drinking water are increasing and if this continues the inhabitants of the city may have to start buying bottled water in the not-too-distant future.

The Baltic Sea is a unique body of water because it is brackish, which means that it is a mix of fresh and salt water. Since brackish water holds less oxygen than fresh water marine animals in the Baltic Sea have had to adjust to an environment containing lower oxygen levels than they usually need. The size of the Baltic Sea is about 400,000 square kilometers, and about 90 million humans live in its catchment area. It is almost a closed sea, with its shoreline circumscribed by Sweden, Finland, Russia, Estonia, Latvia, Lithuania, Poland, Germany, and Denmark. Only beyond Denmark is there a connection to Kattegat, where the waters of the Baltic Sea are occasionally renewed at the connection between the Baltic and the North Sea. This is another critical aspect of the water in the Baltic: it only renews approximately every 30 years. Thus, what comes into the Baltic stays there for a long time.[2]

Moreover, the deep water in the Baltic does not re-oxygenate by itself, but is re-oxygenated by an inflow of oxygen-rich salt water from Kattegat. Large inflows are unusual and happen approximately once every decade.[3] With these inflows, additional salt water is added to the waterbody, and although this water is rich in oxygen, it is causing further complications. This is because salt water is heavier than fresh water, creating a situation with oxygen-rich light fresh water on top of oxygen-poor heavy salt water. Due to this divide—between salt and fresh water—the more oxygen rich fresh water does not mix with the oxygen-poor salt water. This means that although the fresh water can be oxygenized at the surface, this oxygen does not reach the deep water. The deep water of the Baltic Sea needs these inflows of oxygen-rich salt water, and yet, these inflows simultaneously escalates the situation by dividing salt and fresh water further.

[2] https://balticsea2020.org/oestersjoens-utmaningar.

[3] https://www.smhi.se/en/theme/inflows-to-the-baltic-1.13097.

There are multiple reasons for the Baltic Sea being so polluted.[4] The best known is eutrophication, meaning that the sea is being overfed by nutrients, mainly from agricultural run-off and wastewater plants. This results in a process of intensive growth within the sea. Enormous blooms of algae, to mention one visible example, can turn entire areas of the sea green or red. Under normal conditions algae produce oxygen. However, due to their abnormal quantities, the algae start to "eat" the oxygen of the sea. It is actually not the algae themselves that eat the oxygen, but the decomposing bacteria on the sea floor that consumes dead algae. This causes a reaction whereby the already oxygen-poor brackish water loses more oxygen, leading to large areas of the bottom becoming hypoxic (oxygen-poor) or anoxic (without oxygen)—in other words, the sea bottom becomes *dead*. This is a severe issue for water-living organisms. Despite this, there are some indications that environmental policies regarding reduction of leakage of nutrients into the Baltic Sea have begun to have a positive effect. The Baltic might slowly recover (van Helmond et al. 2020; Vikström and Wikner 2019). However, any recovery will depend on a continuing positive trend, without further harm being done to the sea.

The Baltic Sea has a long history of other types of pollution as well. As mentioned above, for decades, chemicals and metals have been released into the sea by industry and later also from households. One of the most common toxic metal compounds found in the Baltic is tributyltin (TBT), which is used in paints designed for the underside of boats. TBT effectively kills the organisms that can damage boats, but it is just as effective in killing non-target organisms. Another metal with a high presence in the Baltic is lead, whose primary source is the petrol used in cars and boats. Trash scuba divers have discovered another source of lead in the Baltic: leaking car batteries dumped in the water, something we will consider in greater detail later. From the 1970s onwards, the regulation of chemicals and metals leaking into the Baltic Sea has improved the situation. However, new types of pollutants continue to be disposed of in the marine environment.[5]

[4] https://www.wwf.se/hav-och-fiske/ostersjon/unikt-innanhav/; https://www.siwi.org/publications/water-pollution-data-baltic-sea-basin-local-regional-approach/ .

[5] https://www.helcom.fi/wp-content/uploads/2019/08/BSEP122-1.pdf; https://www.havet.nu/miljogifter- .

Regular waste, such as plastics, scraps of metal, and rubber, represents another type of pollution. Earlier, this waste disappeared from sight, since there was not enough to resurface and call attention to itself. Today the story is much different: trash floats up all along the shore and marine animals die from eating trash or being entangled in it.[6] In Sweden, the view of water as a suitable waste dump has changed drastically. In 1964 the film *Sjövett* (Sea Manners) appeared on Swedish television. In this film, the public was instructed on how to dump waste in the water.[7] Later on there has been a more aware but still not fully enlightened approach. The 2018 film *Ett badkar kommer lastat …* (A Bathtub Comes Loaded With…) by the Swedish Sports department's national cleaning organization *Städa Sverige* (Clean Sweden) is one example.[8] It shows how no less than the equivalent of 43,800 bathtubs full of trash floats onto the west coast of Sweden yearly. Here one can recognize the extent of marine pollution from anthropogenic trash. However, the Swedish role in polluting water is still not fully acknowledged. Throughout my fieldwork, I have spoken with people about the old practice of deliberately disposing of waste in the water. At first, most people were shocked at the thought, but eventually many could remember how "when they were young and did not know any better" they went out in small boats and dumped old washing machines or other home electronics. There were also those who recalled parking their old car on the ice during the winter so that it would sink once the ice melted in the spring. Sweden's role as a polluter clearly needs to be addressed. The pollution has historically been on both a governmental and an individual level, although individual practices have followed governmental encouragement as to how to dispose of waste. Even though current practices might be better, there is still a historical legacy of disposing waste in the Swedish water ways. That waste is now resurfacing or causing problems in the underwater world.

The situation in Lake Mälaren is similar to that in the Baltic Sea, primarily with regards to eutrophication, chemicals, metals, and waste. Not too long ago, the Baltic Sea and Lake Mälaren were one body of water, but around the year 1100 Lake Mälaren was separated from the Baltic Sea due to the land rising. Now these two bodies of water are connected via three floodgates located in Söderström, Södertälje, and Hammarbyleden.

[6] https://www.havet.nu/?d=3485.
[7] https://www.youtube.com/watch?v=XrcrX9Qaw2o.
[8] https://www.youtube.com/watch?v=eFLZgG0-8NU.

Lake Mälaren is Sweden's third-largest lake, with a surface area of 1140 square kilometers. Increasing levels of eutrophication continue to be a significant problem. This is due to new developments in agriculture, lower water levels, and the elevated volume discharged from human wastewater plants. Lakes are shallower than oceans, and therefore more light penetrates the water and makes it warmer—factors that escalate the growth of plants.[9] More plants mean more biota that need to be processed at the bottom, resulting in more decomposing bacteria consuming oxygen.

Lake Mälaren has a long problematic relationship with pollution. In years past, most pollution came from industry and/or agriculture. However, these sources have been managed rather effectively by policy regulations. Today the sources of continuous pollution are more diffuse. Pollution now comes from chemicals being flushed down toilets and from car tires that wear down, resulting in large amounts of microplastics containing toxic chemical compounds that eventually leak out into the environment and end up in the water. Thus, even if many policies are in place to curb pollution, there are chemicals, metals, and organic pollutants that will still find their way into Mälaren.[10] The lake stores many polluting elements in its sediment as well,[11] which leak out over time, and may bioaccumulate inside fish, birds, and other marine organisms.

Marine life in Lake Mälaren has long been affected by this pollution. Studies have shown high levels of a number of damaging pollutants in fish from the lake, and levels often increase closer to Stockholm. The most alarming recent finding (for human concerns) is in a study by the Swedish Environmental Protection agency, showing high levels of per- and polyfluorinated substances (PFAS) in fish, leading to the recommendation that adults only eat fish from the lake once or twice a year, and children not at all.[12] There are other types of pollution than PFAS in the water as well, such as mercury, medicines, plastics (more in Chap. 4), and lead (more in Chap. 6), among others. And since wastewater treatment plants cannot filter out all sorts of pollutants coming from human societies, the mix of different components is changing by the day as new products are

[9] http://www.malaren.org/malaren/hur-mar-malaren/overgodning/.

[10] http://www.malaren.org/malaren/hur-mar-malaren/miljogifter/.

[11] http://www.malaren.org/malaren/hur-mar-malaren/miljogifter/sedimentprovtagning/.

[12] https://www.svt.se/nyheter/lokalt/sodertalje/ny-rapport-hoga-halter-miljogift-i-fisk-fran-malaren-klart-man-blir-orolig#:~:text=-.%20En%20vuxen%20bör%20bara%20äta,säger%20Cecilia%20Hedfors%20på%20Naturskyddsföreningen.

developed and used in society. Or when, for example, banned products reaper in new shapes, like when a number of PFAS sources were prohibited, new and very similar versions appeared on the market and in the environment, with almost identical environmental downsides. This mass of pollutants and sources is making the tracking of the effects on marine and human life difficult. Since, even if we manage to track levels of one chemical, for example, it is hard to know how it can/will react in combination with other chemicals and environmental circumstances.

One thing is clear, however, and that is that when one reads about pollution in relation to fish in Lake Mälaren, the discussion mainly focuses on toxicity levels with regards to human consumption of this fish (e.g., pike, perch, or trout). This is similar to how discussions about cod being threatened by eutrophication in the Baltic Sea are framed—as soon not available for human consumption. Hence, levels of toxicity in fish, and their stock numbers, are measured with human consumption as the implicit baseline.

Underwater Ethnography: Methodological Strategies

Water is not an element in which humans exist. We, as a species, are not made to be underwater for an extended period of time. However, humans have always been fascinated by, lived beside, and depended on water. As Andrea Ballestero (2019a: 406) has written, "Water is always more than itself; it is a political field where people elucidate what it means to live life collectively in a world that is always more than human." She urges scholars of water to avoid taking what water "is" for granted, as the materiality and ontology of water is context dependent and changing. So, how do we study the underwater world with the methods available to anthropology?

Water as a field of study is not new to the discipline and several recent studies have shed light on many aspects of life with and under water.[13] For example, Jessica Barnes (2014) used a political ecology lens to follow the movement of water from the Nile into Egyptian pipes and fields, across national borders, and in interaction with human and nonhuman life. Furthermore, she explores scientists' technical approaches as part of nation

[13] For anthropological work on water see Ballestero (2019a, 2019b); Barnes (2014); Barnes and Alatout (2012); Hastrup and Hastrup (2015); Helmreich (2009, 2011, 2023); Krause (2013, 2017); Krause and Strang (2016); Lien (2015); Strang (2019, 2023); Schmidt (2019); Orlove and Caton (2010).

building that centers around access to water (Barnes 2016). Another common approach in the anthropology of water has been to investigate the linkage between water and infrastructure. Lisa Björkman's 2015 study of Mumbai shows the complicated social relationships that spring from a (mal)functioning water supply infrastructure. There is also the field of maritime anthropology that has looked at fishing communities (McCormack and Forde 2020). Shankar Aswani (2020) offers a current comprehensive overview of the field, arguing for the need to seek new collaborations to help protect the communities under study due to contemporary socio-economic and environmental challenges.

One strand of the anthropology of water that is of especial interest to this **book** is that which shows how water makes and is made by and with bodies. These scholars show that water can transport pollutants and waste over time and space, and how it interacts with the world around it, human and other-than-human. Such studies also emphasize the arbitrary situation with regard to responsibility, showing that water pollution is hard to prove juridically, hence making it difficult to charge the polluter responsible (Bakker 2012; Hamdy 2008; Hoover 2017; Montoya 2017; Wolfe 2017). However, when individuals raise their awareness of water, they may call on the state to provide the basic service of keeping water clean and accessible (Paerregaard et al. 2016; Anand 2017). Or as Ballestero (2019a: 410) concludes, "When people find themselves in deeply antagonistic relations with the state, they craft nonnormative subject positions such as that of a water defender or water protector." As will be shown below, this subject position defines both the scuba divers and the scientists among whom I have worked.

In the studies cited above, water has mainly been studied through relations that can be seen or tracked from land.[14] I seek other ways of engaging with water. Stefan Helmreich (2009) offers one such approach when he joins marine scientists on a research vessel in order to study microbes in the sea. Helmreich does occasionally go below the surface himself using a submersible (a kind of submarine). In this way, he approaches the study of water with a method of looking at water from beneath the surface. Nevertheless, he is still not *in* the water; he observes from behind a glass window, from within the submersible. However, water is a rich sensorial element that gives and takes life, diffuses heat 20 times faster than air,

[14] For two examples of human-water relations with a focus on bodily emersion, as in swimming, see Bonnie Tsui (2020) and Eric Chaline (2017).

transmits sound differently, the visibility of which varies extensively, and, not least, is home to countless other-than-humans.[15] In my work, I have accompanied scholars such as Stephanie Merchant (2011), Justin Raycraft (2020), and Nils Bubandt (2023) in seeking to engage with water differently, by applying embodied research to know water's relational and sensorial aspects. I do this by joining a group of scuba divers and submerging myself into the environment under study. Further, I scrutinize the bottom sediments of the sea with the help of marine scientists who use various technological devices to extract sample cores from the bottom and bring them up to the surface where they can be examined hands-on and analyzed. These scientists also use different types of sonar, take water samples, and more.

The following ethnographic study was conducted from spring 2019 until early 2024. It is a multi-sited investigation that pursues an idea and an issue, rather than focusing only on a particular site (Marcus 1995). As is often the case with multi-sited fieldwork, some places are more central than others. In my case, the main sites have been Lake Mälaren and an area of the Baltic Sea, in the vicinity of Stockholm. Even so, the **book** ends with a comparison of underwater worlds on different continents, highlighting how water differs materially between places, while showing that the issues of waste and pollution remain similar. As Pauwelussen and Verschoor (2017: 298) have written, "Shared worlds are not necessarily place, or land-based at all." There are multiple underwater worlds, all different, yet interconnected and affecting one another.

In this book, I am mainly following marine scientists and trash scuba divers in their attempts to engage with and assist the marine world. This does not mean that there are not numerous other actors that are important (government officials, other environmental organizations, etc.), and I do engage some of them briefly throughout the book, but I have made the

[15] The term "other-than-human" is relatively recent, coming out of the multispecies turn in anthropology (Kirksey and Helmreich 2010). However, this representation of and engagement with animals and other beings are not new to the discipline (Lien and Pálsson 2021). For example, in *The Nuer* by Evans-Pritchard from 1940, animals are given due credit as parts of social life. Despite this long tradition of engaging the other-than-human, it is now being done in a different way. If other-than-humans were formerly part of the social dynamic needed to understand a people, they are now increasingly becoming the topic of study in and of themselves (Schulthies 2021; Olsen 2021). Moreover, scholars have begun looking into multispecies justice (Chao et al. 2022), a response to human treatment of the world and beings in it.

decision to focus my writing on the marine scientists and the trash scuba divers because I view them as especially important for the pro-action and pro-change story I hope to tell. Furthermore, I have decided to keep my narrative around the work closest related to water. For example, with the trash scuba divers I have followed how waste has been brought up out of the water, and how the divers interact with it, and then bring it into the waste management system. But here my description ends. It would have been interesting to also venture into the waste management system, but it would take me a bit too far away from the key story I am focusing on, not to mention that waste dumps have already been carefully studied by others (e.g., Reno 2016; Millar 2018; O'Hare 2022). It is not possible to follow all threads and actors that this study enlivens, but hopefully through these necessary cuts, the story will become more decisive.

In what follows, I will describe different aspects of the project, elaborate on the methods used, and reflect on my positionality within the different groups. I will also address the profound impact Covid-19 has had on my fieldsites and interlocutors. Additionally, it is important to note that I worked with the groups of divers and scientists simultaneously, even if they are described separately. The methodological section ends with a reflection on visual methods.

Engaging with Trash Scuba Divers

When I applied to do this research project as a PhD student at the Department of Social Anthropology at Stockholm University, I was already in contact with a group of trash scuba divers.[16] However, as I was about to begin my work, I realized that this group only conducted cleanup dives once a month, and only during the warmer part of the year. This seemed like a problem because they did not meet regularly enough for immersive field research. This hardly felt like anything remotely close to what I had

[16] Trash scuba divers is a popular denomination for scuba divers using an open circuit regulator system who dive to collect debris from the ocean floor. Open circuit means that the breathing gas is released from the system, one breath at a time. Closed systems retain the gas and therefore do not release any bubbles. These trash scuba divers would otherwise most likely be conducting recreational diving as a form of leisure activity. Some trash scuba divers do use a closed-circuit system, but mainly as a promotional exercise, rather than for any actual gain from that more expensive, advanced system. A closed-circuit is otherwise employed for deep and technical diving or in the military. For a discussion of technical deep diving, see Brown (2020).

imagined when deciding to follow the work and impact of local trash scuba divers in Stockholm. This caused me to seek other groups who were doing something similar that might be a better fit, and I did find one (Fig. 1.2).

I discovered a diver who was collaborating with a few individuals and had begun cleaning waste from the waters around central Stockholm during the spring of 2019. He planned on diving multiple times a week, and no less than once a week, all year round. I contacted the diver in question and we met during one cleanup. We got along well. Moreover, as he was in the middle of recruiting new divers for the project, he was happy to learn that I had been a diver for over a decade. We agreed that I could follow the project, see how it developed and the work accomplished, and in return I would take an active part in the work. This seemed like a perfect opportunity for ***participant observation***. The initial group consisted of two regular divers and a few persons active on land. From 2019 until mid-2022 the group grew to over 300 members who joined in cleanup activities on a weekly basis, and it continues to grow by the day. I have taken part in a minimum of one, often more, cleanups per week between 2019 and 2024,

Fig. 1.2 This image depicts a trash scuba diver, a car tire, and sediment in the water. (Photo: Björn Alvarsson)

having dived on over 150 of them, with dive times between 40 and 90 minutes per dive.[17] On the other cleanups where I have not acted as a diver, I have functioned as a "rope puller," the person standing on land, holding a line connected to the diver. The line is also used to pull objects out of the water, as illustrated in the introductory vignette. In the second part of the **book**, I ethnographically describe the work of divers and rope pullers in greater detail. Furthermore, I will elaborate on the cyborgian aspect of diving (Jue 2020; Raycraft 2020), namely, how technology allows a terrestrial body to submerge into the aquatic environment for an extended period of time.

The role of technology is of great importance in understanding both the practice of scuba diving and the way knowledge is produced in scientific laboratories. Scuba divers and marine scientists both depend on advanced technical devices that allow them to enter and/or study underwater worlds. They use a variety of technical devices and equipment that allow them to see different things, things that would not be accessible otherwise. This raises fundamental epistemological questions discussed among others by Science and Technology Scholars (STS) such as Bruno Latour and Steve Woolgar (1979), feminist scholars like Emily Martin (1996), and anthropologists like Cymene Howe (2019).

In the course of my work I have conducted numerous semi-structured interviews. These were mainly in Swedish and have been translated to English by me. However, most of the material I collected is in the form of unstructured interviews during, around, or when planning the cleanups. The cleanups themselves are intense, and people rarely stand around to conduct formal interviews, although participants are happy to talk while working (Driessen and Jansen 2013; Davies 2008; Hammersley and Atkinson 2007). In due time, the group organized itself into a formal non-profit organization. I have been following this work closely, and have even been included on the organization's board as a secretary. Since, as the leading diver told me, "You go around writing down everything anyway. You might as well be of some use while doing so." To be open about my role as a researcher, but also willing to lend a hand in the work, was the

[17] One point regarding the number of dives is interesting to note. From the time a diver does their first dive until they can reach a professional level of diving (Dive Master) as recognized by the world's largest association for divers PADI, the minimum number of dives is 60. Hence, it is possible to argue that I have dived enough to become a professional trash diver during this project.

only viable approach with this group. There was no interest in having a hang-around researcher, it was an option between participating or not being there at all. I want to state that I, as a person, believe in and am committed to the work of the group. However, as a researcher, this did not prevent me from reflecting critically on the practices and events that unfolded during fieldwork. To be active in the group you conduct research with is a path taken by others before me.[18] For example, Kim Fortun (2001) was actively contributing to the work of her interlocutors in battling powerful companies in the aftermath of the Bhopal disaster in India; organizing events, writing petitions, and more. Ömür Harmanşah (2020: 49–50) argues for what he calls "fieldwork activism," writing: "My sense is that our moment is one of activism, and it is one in which theory will emerge from the field." My active involvement with the diver group has allowed me to be part of certain processes that would not have been possible by observing as an outsider. For example, as will be elaborated in Chap. 7, the divers knew that I work with scientists that study plastics in the marine environment. So, when the group planned a large exhibition with a special focus on car tires, they asked me to contact those scientists and ask if they could provide information that the divers could use in their presentation. I became the link in the relationship between scientists and environmental activists, thanks to my active involvement with both sides.

The fact that I myself am a diver has been crucial in enabling me to understand the diver's point of view: the environmental, psychological, and sensuous concerns during trash dives. Or as one of the pioneers of modern-day scuba diving, Jacques Cousteau, put it in 1956: have the "flesh feeling what the fish scales know" (quoted in Jue 2020: 45).[19] However, as the introductory vignette illustrates, these trash dives are no

[18] Dan Podjed and Meta Gorup (2021: 5) argue with reference to Gupta and Ferguson (1997) in an edited volume named, *Why the World Needs Anthropologists*, that it is time to "dismantle the notion of 'applied' anthropology as the 'evil twin'" when arguing for a move for anthropologists from mere observers and critics to an active engagement in and with the topics under study. Anthropology offers a unique perspective on social issues, and in a time of environmental crisis (among other issues), this perspective can turn out to be crucial. Thus, in my work I attempt to put the methods and theories of anthropology to use in active involvement with the groups among whom I work. In order to not only "Do No Harm" but to also "Do Some Good" (Podjed and Gorup 2021: 7, referencing Briody and Meerwarth Pester 2014: 30).

[19] Jacques Cousteau was a French naval officer, divemaster, and filmmaker who co-invented the first open-circuit SCUBA technology. Among his further accomplishments include the first underwater documentary, and he also wrote a series of books about the underwater world. The most well-known is *The Silent World: A Story of Undersea Discovery and Adventure* (1953).

regular dives. Such situations test the diver on many levels. Above all else is the mental capacity not to panic when visibility disappears. Embodied diving skills, allowing a diver to maintain good buoyancy even during these situations, are equally important. A diver must learn how to react during these situations, or one could be in danger (Dimmock 2009; Dimmock and Wilson 2011). These dangers of trash diving lead to the next point: being a diver *while* researching divers is key, that is, the sensuous experience of diving can best be understood through embodied and emplaced research.[20] The researcher will learn about bodily practices with their body while situated within the environment, time, and place of the study, that is, being emplaced (Pink 2015: 28). This is illustrated by Yolanda van Ede (2009: 62) when she writes about the importance of sensuous anthropology:

> The senses are not taken as merely a topic of research, distanced and objectified from the researcher to investigate, but an inherent part of consciousness-raising process that requires a broader methodological touch; more than observing and interviewing alone.

This is especially true in the cold dark waters encountered by trash scuba divers in Sweden. If one does not apply an embodied, emplaced, sensuous approach, it is hardly possible to make sense of this type of diving experience and learn how tactile obstacles impact the work being done. "We learn how to see; we learn to observe or not to observe. This is the same for all other senses as well" (van Ede 2009: 65). Hence, to learn what it means to dive without visibility, to "feel" your way around, is paramount if one is to engage the field sensuously, and absorb the skills and practices as one learns them (Grasseni 2007).

I had initially planned to dive behind other divers, to observe, and later describe the way they worked. With the low visibility in these waters, this turned out to be impossible. Not only could I not see anything following a diver, but I would also make the situation harder for that diver by decreasing the visibility even further (I elaborate in Chap. 6). Hence, to ethnographically understand and describe the practice of trash diving, I have had to adopt an auto-ethnographic approach (Reed-Danahay 2019),

[20] For embodied methods cf. Merchant (2011); Pink (2015); Rodineliussen (2017, 2019a, 2019b, 2021); Wacquant (2004); and Wulff (2008).

using myself as the interlocutor from the water and bringing the reader along with my sensorial experience of trash diving.

Early into my fieldwork, the world was struck by the Covid-19 pandemic and most group activities shut down. Fortunately for me, the cleanups continued, because it is an outdoor activity, hence it was possible to keep the recommended distance between people. The pandemic had another interesting turn for my project. With most activities closed down, people got restless, and sought a social activity that would allow them to meet other people and do something physical. These cleanups offered both. It was evident that after Covid-19 restrictions came into place, the number of volunteers for the project increased dramatically, as discussed in Chaps. 6 and 7.

This **book** also includes material from Brazil and Thailand. The decision to include a comparison between countries and water came from my initial experiences when diving in the dark waters around Stockholm. When I began this fieldwork in Sweden, I had mostly been diving in water with better visibility, and had spent much of my time as a diver in Thailand and other countries with high-visibility waters. I realized, that in order to emphasize how little visibility there is around Stockholm, to make readers really grasp the added stress on divers in these waters, a comparison was needed. I decided to use Thailand as one example, due to my earlier experiences from the country. To freshen up my understandings of Thai waters I went on a one-month fieldtrip to Koh Samui, winter 2019. Although the trip was rather short, it is informed by my longer experience from Thailand, where I have been diving from Thai islands for five to six months over a decade long period. Furthermore, I also make use of the work I did for my master's thesis in Brazil from 2017, adding another place to the story. Thus, the comparison follows my own trajectory as a diver and a scholar. Both cases are included here as comparative studies to illustrate ethnographically that the issue at stake, politicians' responses and promises (that are sometimes hollow), and divers' engagement, have many similarities with the Swedish case. Two particular differences between the water in Brazil and Thailand compared to the Baltic Sea and Lake Mälaren are visibility and temperature. In Brazil and Thailand, the visibility extends much further, making trash diving easier, but also changing the places where waste has been dumped to more remote locations. Temperature is another key factor: diving in colder climates becomes harder and more technical as additional gear is needed to protect the diver against the environment. The comparison is also important because, as I will describe in Chap. 8

and further in the concluding chapter, the trash scuba divers have begun an expansion seeking to reach outside of Sweden. This will lead them to other types of water and other experiences with visibility. Understanding the difference between types of water is one key aspect to understanding the practice of trash scuba diving, but also to what type of waste water can hide, and where.

Fieldworking with Scientists

The Covid-19 pandemic profoundly impacted the way I could conduct fieldwork with "the scientists" (different groups, with different disciplines). I followed their work on research expeditions at sea and in laboratories. I went on one expedition in early June 2019, and another in late August of the same year. Also in that year I joined a trip to Stockholm University's field laboratory on the island of Askö in the Stockholm Archipelago. I had planned at least three expeditions for spring 2020, but they were all canceled due to the pandemic. It turned out that it was not possible for non-essential persons to join ongoing expeditions, and a social scientist did not satisfy the definition of "essential"; therefore, the pandemic severely reduced the time spent with scientists in the field. Luckily, however, I did manage to visit three different laboratories before the pandemic struck. I followed samples collected on the June expedition to two laboratories in the Netherlands, in Utrecht and Nijmegen, in the autumn of 2019 (Fig. 1.3). I was also able to go to another laboratory in Trondheim, Norway, in February 2020 where scientists were working on plastics in aquatic environments.

My interaction with the various scientists did not end with the on-site visits. I remained in close contact over the phone, through messages, and via emails with several of them. They have invited me to online lectures and seminars and have shared their writings—some of which I have included as references throughout the **book.** There is only one article based on the expedition I was part of mentioned, but that is because it is the only one yet published. Other references to the involved researchers are from earlier expeditions, but on similar material. More recent publications from the Norwegian scientists are included. And, as mentioned above, I have also followed a collaboration between the scientists from Trondheim and the divers in Stockholm (explained in Chap. 7). Therefore, one could argue that my research among the scientists has been hybrid,

Fig. 1.3 This image shows a core sample from the bottom of the Baltic Sea. (Photo: Author)

with intense in-person visits followed by a longer online engagement. By means of these hybrid methods, I was able to adapt to the pandemic.

There are a number of scholars who have investigated ways of conducting ethnographic research online from whom I have drawn inspiration. For example, Sara Pink et al. (2015) has written about how to conduct digital ethnography and about the role of online practices in the world today. I have used Tetyana Lokot's (2018) ideas about hybrid online practices to think methodologically. Nowadays, relations take place online, and so does the dissemination of knowledge and research results. Hence, it is important to also engage with these practices via the same platforms upon which they operate. For this reason, I argue that combining online and offline methods is more than me adapting to a complicated situation for research during the pandemic, it is also a way of adapting my research to the world of today more generally (Geismar and Knox 2021; Hjort et al. 2017).

In Chap. 2, I will further elaborate on how I gained access to the field among scientists. It was not straightforward, and in all honesty, it came down to luck. I have reflected quite a bit on my positionality in this field. Whereas I conduct fieldwork sideways among the scuba divers, it felt like I did a more upward-ish fieldwork with the scientists (Hannerz 2006, 2010). I am not referring to the relationship between the social and natural sciences, but to academic seniority. This led to, in various degrees, that the tone could sometimes become a bit didactic when scientists discussed their research topics. This was even more pronounced with senior researchers as they spoke about their area of expertise in an authoritative way. As I listened and observed their work, I learned about their relationship to the water and the other-than-humans within. My interest was not to learn of their findings per se, but to analyze what they had done—methodologically and relationally—to engage with and understand the underwater world.

I conducted interviews during in-person visits, then followed up with mainly unstructured interviews and work-alongs. On the research vessel and in the laboratory, we would talk while working. I would then pose follow-up questions and conduct interviews or have unstructured conversations on the topic of their research online. I have also joined several "Baltic Breakfast" seminars hosted by the Baltic Science Center at Stockholm University (the center runs the research vessels). In this way I kept up with ongoing science with impact on the Baltic Sea. At first, these breakfasts were in person, but with the pandemic, they too moved online.

The fact that I was not directly concerned with the scientists' research findings, but more in how they related to their work and the undersea world helped me to gain and keep access to the field. In my visits to the laboratories and discussions with heads of research teams, their most common concern was the risk that their research findings would leak out through me prior to getting published (some of the scientists had previous experience of findings being "stolen"). The solution was that I promised to let some of the researchers look over my writing covering these visits to scan for possible research findings that should not be there. It was nervous to send my writing to the researchers, as it was my understanding and description of their everyday work they would be assessing. However, the reactions were positive. I was told that it was fun to read a kind of diary from the work-day. But more importantly, it solidified our relationship. One of the researchers said:

> When we are on the research vessel, we all have our different tasks and goals. But when reading your descriptions and analyses of the relations that bind us together, it became clearer to me how much our work really is dependent upon each other. It is important to remember that even if we each work with detailed aspects of the world, they are still part of the same world.

This reflection is a strong argument for the importance of a holistic anthropological project. It is by finding the relations between different but interlinked life-worlds that we can first understand and then change the way the world works.

Visual Methods

I have taken great interest in visual aspects during fieldwork, both as a method and as material for analysis. There are at least two ways of engaging visual material in research: *producing* and *collecting*. I produced a significant number of photographs and videos during my research, capturing moments on the water, in laboratories, on land, in wastewater treatment plants, and underwater. I mainly use my smartphone camera for surface and a GoPro for underwater photography. When producing visual material, it is essential to consider the question of framing. That process is deeply political as it determines what will be saved for a later time (Butler 2010; Hannerz 2004; McDougall 2005; Rodineliussen 2019c). Moreover, the selection involved in framing has increasingly become part of a negotiation between the picture-taker and the individual being photographed. This is due to the increasing awareness among the public as to how images move around and *do* things. Sometimes this is unintentional, and at other times there is a clear agenda shaping the way we view the world (Favero 2007, 2020; Collins et al. 2017; Wright 2008).

When collecting visual material, I have mainly followed social and regular media channels that have reported on events related to my fieldwork. These images, and sometimes video or live TV reports, have become ethnographic descriptions of the field. I have also used these materials in photo-elicitations with interlocutors (Rose 2007; Maaker 2000; Larcher and Oxley 2015). These elicitations have, on many occasions, happened in one-to-one meetings with interlocutors. However, I have also conducted group elicitations. I have had a group-messenger chat with the diver group, and their images taken during cleanups have come up for discussion. This has been a sort of online-asynchronous photo-elicitation where interlocutors have talked about their relationship to a specific image, at a time that fitted them. These comments have not been anonymous; rather they have functioned more to collectively formulate a group perspective on a particular image. The participants sometimes joked about an image as something weird, and at other times rallied behind it, urging it to be posted online to the group's social media accounts as it contained something that the group wanted to support.

The diver project has yielded plentiful images. There have been several professional photographers documenting the work. Most of the group's social media images were taken by them. These photographers have various styles and different viewpoints regarding both how to take images and what to photograph. I have talked quite a bit with them about this practice in relation to the field. Many images in this **book** are theirs and are used by permission, crediting the photographers (permission have also been given by them credited as anonymous). As taking photos of the work is their main contribution to the diving project, I seek to use their images when I can to both acknowledge their contribution and share their perspectives on events.

Aquabiopolitics: Management of Waste and Marine Life in a Water Assemblage

As mentioned earlier, the term *aquabiopolitics* is based on Michel Foucault's notion of biopolitics. Biopolitics, or biopower as he often called it in his early writings, was first mentioned by Foucault in *The History of Sexuality* (1978) where it refers to the reproductive qualities of individuals as part of a politics of human bodies. Then, in his 1977–1978 lecture series at Collège de France entitled *Security, Territory, Population* (Sennelart 2009), biopower is termed biopolitics, and the focus increasingly turns toward another of Foucault's concepts, "governmentality," as a term to analyze the management of human populations. This leads to Foucault seeming to lose interest in the idea of biopolitics (Wallenstein 2013: 12), as evident in his subsequent 1978–1979 lecture series *Naissance de la biopolitique* (The birth of biopolitics). There biopolitics is only mentioned briefly in the introductory lecture and then in the concluding remarks, although it is the stated topic of the series (Holmbäck and Wallenstein 2014). However, as Johanna Oksala (2013) explains, Foucault's turn toward neoliberalism and governmentality is not a move away from biopolitics per se, but a direction that is essential for a grounded understanding of biopolitical logic at play as a form of governance in what Foucault sees as a neoliberal world. Thus, for my purposes in this **book**, governmentality and biopolitics will be seen as closely interlinked.

However, aquabiopolitics departs from Foucault's human-centered understanding of biopolitics. Here I follow Thomas Lemke (2013), who shows that

> the reluctance to engage in a purely negative critique often leads to a complete lack of an evaluative perspective, and therefore to a "technical" theory that duplicates its object of study; inversely, governmentality studies have been reluctant to integrate analyses of technical and non-human networks. (as summarized by Wallenstein 2013: 30)

Therefore, while I build on Foucault's ideas about biopolitics as a way of governing human populations, I take a cue from Lemke (2013) and propose that nonhuman populations also ought to be included in our analysis. I will do this by following what has been argued by Nicole Shurkin (2009) and ethnographically exemplified by Emily Wanderer (2020) in her work among Mexican scientists who deal with so-called invasive species. They hold that biopolitics also extend into the realm of other-than-humans, following capitalist logics that impose themselves on nonhuman bodies. Thus, my understanding of the "bio" in Foucault's biopolitics is in this more inclusive way, emphasizing that human life is produced through ways of governing over other-than-human life. I hold this understanding in the company of a number of scholars. For instance, Bruno Latour extended Foucault's use of biopolitics in 1988 when he applied it to microbiological activity. In so doing he added the prefix "micro," and coined the term microbiopolitics. Heather Paxson has developed this further in relation to microbial work in the making of cheese (2008, 2012). Paxson writes that the "neglect of the microbe (any organism, in the singular, invisible to the naked human eye) continues to distort our anthropological view of the social world" (2008: 18–19). She emphasizes the importance of understanding microbes as a "companion species" (Haraway 2003). No matter how invisible microbes have been, and still often are, they play a crucial role in the human becoming. Paxson (2008: 18–19) argues that the concept of microbiopolitics "concerns the recognition and management, governmental and grassroots, of human encounters with the vital organismic agencies of bacteria, viruses, and fungi." For my purposes, Stefan Helmreich's (2009: 4) application of microbiopolitics in relation to marine life is of special importance as he helps me to turn the gaze toward the marine environment. Helmreich aims to show "how marine microbiologists seek to make new scientific knowledge about biological life forms meaningful to social and cultural forms of life."

In the pages that follow, I write and think about water as an assemblage wherein multiple lifeforms jointly constitute life in relation to one another as an "open ended gathering" in a mutual, but not equal or symbiotic, process of continuous becoming (Tsing 2015: 22–23). I follow numerous anthropologists, such as Strang (2019, 2021, 2023), Pauwelussen and Verschoor (2017), Paerregaard and Uimonen (2021), and Lien (2023) in showing that water is more than merely the molecular chemical compound H^2O that is vital to life on planet Earth, according to a western scientific perspective on the world.[21] This is because, when looking at oceans, seas, rivers, or even a drop of rain, what we see is an assemblage of H^2O, chemicals, metals, organisms, and other compounds (or beings). Each drop of water differs in its chemical and organic composition due to these other constituents. Therefore, I concur with scholars such as Linton (2010), Helmreich (2011), Wright (2017), Ballestero (2019a), and Schmidt (2019) in thinking of water more as a process in becoming than a fixed molecular compound. As chemicals, organisms, and animals move around in water, so does the composition of a specific body of water. Thus, the assemblage of water is never static, but always in a process of change.

While microbiopolitics is a helpful term when thinking about microbial life in water, there are also other animals, such as fish, crabs, and seals in water. Moreover, there are chemicals, toxins, oxygen, methane, sulfide, iron, lead, and innumerable other elements present in water, not to mention the humans and other-than-humans living close to or on the water.

[21] I will argue against the common assumption that reduces water to a resource. When looking at the UN's Sustainable Development Goals for 2030, this becomes explicit. Goal number 14, *Life Below Water*, comes with the caption "Conserve and sustainably use the Oceans, Sea and marine *resources* for sustainable development" (UN 2020, italics added). How do we proceed to think about water outside of this politico-economic context? Is it fruitful to do so? I believe it is. We should not abandon this perspective. However, we should also try to think beyond it if we are to engage with water as more than a resource.

For this reason, I find that "micro" lacks the inclusiveness needed to describe the human-waste-pollution-water relationship I am addressing. Helmreich (2014) has built on the microbiopolitics of Paxson (2008), pairing it with the term "symbiogenesis" by evolutionist Lynn Margulis, suggesting the formulation "symbiopolitics." Helmreich (2014: 5) proposes that "symbiopolitics is the politics of living things coexisting, incorporating, and mixing with one another," a more inclusive concept that allows for more than microbes. Nevertheless, the relations I have found in water cannot be encompassed by "symbiotic." I feel that something is still missing.

Kregg Hetherington (2020) has recently proposed "agribiopolitics" to explain the relationship between human and plant health. His approach sees human governmental practices as very selective, ranking different beings and removing those less wanted. It illustrates how different politics make a life or death decisions for a number of plants, with consequences for humans who depend on these plants. In Hetherington's account, a sort of circularity appears: human actions and decisions impact how plants live and die to suit the needs of humans, but in the end, unforeseen effects rebound on human societies. These are very similar processes of relations as those that I have found in my work with life in water. Take the process of eutrophication as described above as one example. Humans overuse nutrients to enhance the yield from agriculture to provide more food. These excess nutrients leach into the water, increasing algae growth that causes a lack of oxygen in the water, thereby damaging marine life, which then becomes a problem for humans who can no longer fish for food as they used to. When effects like these come back to haunt humans, regulations are put in place to curb the problem and improve the situation with human well-being in focus. The intent might not have been to disturb the marine world when using the extra nutrients, but that was the result. Another case is, for example, how drainage around roads adds to pollution of water. When cars drive on the road, tires lose microparticles, these end up in the drainage system, and are washed toward the water when it rains. This is one major contributing factor to microplastics in water, together with dumped tires, and as I elaborate in Chap. 4 and further in Chap. 7, it comes with deadly consequences for marine life. It has now also become a priority for humans, as toxicants travel from roads to water to fish and then to humans when these fish are consumed as food. When it becomes a human problem it also becomes a human concern.

In building on the work of Hetherington, I exchange the "agri" for an "aqua" and add it to biopolitics in order to delimit practices, relations, and effects that unfold within water. However, it is important to note that in Hetherington's usage, politics are focused on the rather conscious efforts of governmental practices and corporate agents. In my usage, political actions are not necessarily initially and consciously directed toward and meant to impact the marine environment negatively, although, they do sometimes have significant impact. This may be hard to see until changes appear on the surface—the result of a form of "slow violence" (Nixon 2011)—and as I will show, it is only when the harm becomes visible that humans start to act.

Furthermore, as Melody Jue (2020), Justin Raycraft (2020), and others have shown, there exists a terrestrial bias in how humans think about water. This means that water is seen as a resource for humans to use. Thus, the term aquabiopolitics seeks to provide an aquatic perspective instead, beginning with the negative results of different practices, rather than the intention behind them in order to highlight that precaution must be taken before, not after, harm is detected. It is important to think one step further ahead—especially when it is not as easy to follow how things develop, as they are not directly visible.

Aquabiopolitics is used as a concept that allows the mapping of different actors who have created the contemporary situation with regard to water, waste, and pollution.[22] In this, not only human agents are recognized but a wider assemblage of beings and substances that constitute the body of water in question. Aquabiopolitics embraces the importance of politics and power while also keeping a relational perspective. Whereas Hetherington begins his analysis of a system of relations from the point of view of on-land observations, I depart from below water and look at the life—or death—along the bottom. I engage with multiple other-than-humans of the marine world, such as microbes, cable bacteria, jellyfish, and worms, and with waste and pollutants in a myriad of forms and levels of toxicity. From this engagement with the underwater world, with waste and marine life, I scale up along these relations, following the circular movement initiated by human practices.

[22] This follows a traditional approach among *political ecology* scholars in looking at and mapping relations of power; cf. Karlsson (2015); Barnes (2014); Robins (2020); Paulson and Gezon (2004).

In my understanding of water as an assemblage of multiple relations that become together, I find Hannah Knox's (2020: 7) explanation of such a perspective useful. She writes, following Eduardo Kohn (2013):

> An anthropology beyond the human is not a posthuman anthropology but an attempt to extend anthropology's remit to be able to attend to representational capacities that the modern social sciences have tended to bracket out as not central to human meaning-making-processes.

Knox's argument allows me to extend the idea of "water as a total social fact" to also include the other-than-humans of the sea (Orlove and Caton 2010). For example, in examining wetlands in Turkey, Caterina Scaramelli (2021) shows that environmental politics are not merely directed at controlling other-than-humans, but are made through them. Similarly, the myriad other-than-humans of the sea can be viewed as embodied agents of human politics, past, current, and future.[23] This is further exemplified by Stephanie Wakefield and Bruce Braun (2019: 198), who write about human-oyster relations in managing extreme flooding in New York. They describe oysters as part of a living infrastructure: "oystertecture," illustrating how managing the life of oysters has a direct impact on human life. This leads, as they put it, to "an instance of biopolitical doubling, we now manage other life to secure human life."

Maria Puig de la Bellacasa (2017: 130, 155) has argued for an "alter-biopolitics" to avoid reducing biopolitics to preserving human life. She advocates an engagement with ethics, thinking "of the ethical as a social practice, as a living technology with material implications in remaking human and nonhuman ontologies." I find her ideas illuminating and crucial for thinking of life writ large in other-than-human and human worlds—especially in how she puts a relational engagement and understanding in the forefront. While there is undeniable value in "de-subjectifying" the human, I find it hard to do so in the story I will tell in this book. In my analysis of the human-water relationship, it is problematic and, in a way, immoral not to make the human role explicit. However, although Puig de

[23] Throughout this book I will engage with issues of time. We will ponder the concept of *deep time*, originally a term from archeology and geography meant to describe earth processes over vast timescales. We will think about how slow processes and long-time results of pollution construct the world of today and tomorrow, for generations to come (Ialenti 2020; Irvine 2020), mainly thinking along the lines of "slow violence" as per Rob Nixon's (2011) definition.

la Bellacasa (2017: 162–164) seeks to "de-subjectify" the human, she also concludes that caring behavior is a situated practice and therefore has to be understood within a specific context.[24] In some situations, humans (for example) can find themselves in a position where they must take a more active role in caring for the world for the sake of humans and other-than-humans, and "in some contexts, care is inseparable from killing" (Puig de la Bellacasa 2017: 164). This leads us to the aquabiopolitical challenge of how humans can care for humans by the management of marine life.

As noted earlier, I use aquabiopolitics following Paxson's (2008) argument for microbiopolitics as they highlight both the recognition and the management of human encounters with the other-than-human. Aquabiopolitics allows an engagement with scuba divers who question human practices of disposing of waste and pollutants in water, using water as a secret dump where things "disappear" because they are out of sight. These divers work to unveil such practices as part of managing life below water. The term also allows an engagement with marine scientists who take samples from the water to map pollution patterns, find other-than-humans that can work with humans to battle the pollution, and seek to understand what can be done to lessen the ongoing damaging relations between societies and water.

Aquabiopolitics as a term connects the two parts of my research—scuba divers and scientists—and contributes to focus in on questions in each case toward opening up the underwater world: making the invisible visible, and thus, knowable and relatable. Moreover, with the increasing visibility of the correlation between human societies and underwater worlds, also relationships of power and governance become evident. In my understanding of aquabiopolitics, humans are those who govern on the beings below water, with the ultimate aim of enhancing life for humans on land. This governance is at times explicit, and at other times more implicit, but as I will show, it exists and affects nonetheless.

Thus, throughout the book, I have aquabiopolitics as a template for understanding how the situation I describe has become what it is today. That includes relationships that have created one of the most polluted seas

[24] I will elaborate on the notion of "care" mainly in Chaps. 2 and 4. I follow the work of Maria Puig de la Bellacasa (2017), Leslie Sharp (2019), and Caterina Scaramelli (2021) in believing that care never can be innocent, and that it always contains some type of violence. The terms care, innocence, and violence will be elaborated in the context of my ethnographic descriptions.

in the world and an increasingly troublesome environmental situation in Lake Mälaren. My usage of the concept is not meant to explain the actions of the scientists or the trash scuba divers. Instead, I use it to understand the reason for their actions, as they seek to undo certain relationships that they view to be wrong and harmful, for humans and not. Aquabiopolitics offers a lens where relations of power can be placed in a larger context that explains the logic behind them. A logic that often comes down to what we define as life, and what life matters the most. Some life appears to be reduced to life without rights, as something that can be killed without consequences for the killer, like Giorgio Agamben's 1988 notion of the Homo Sacer, as a "bare life." A bare life, to Agamben, was a political subject that was alive, but that had no rights to that life. And therefore, could be killed without consequences for the killer. Bare life is thus defined through its exposure to death. Agamben (1998: 4) writes, "In any case, however, the entry of *zoé* into the sphere of the *polis*—the politization of bare life as such—constitutes the decisive event of modernity." Applying this line of thought to the marine world highlights how beings in water—a "new" frontier for human "modern" exploitation (Lien 2023)—can be categorized as bare life when expanding a Foucauldian biopolitical logic. Like Samuel Weber (2012: 7) writes,

> It is reassuring to think that we can make clear-cut distinctions, especially where violence is concerned. It is reassuring because we like to believe we can clearly and unequivocally distinguish violence from nonviolence, destruction from construction, killing from vivifying, death from life. It is reassuring not least of all because such clear-cut distinctions seem to give us a hold on whatever is being distinguished while keeping us at a safe distance from it.

Marine life gains importance if it is important for human life, and if not, then not. Humans care for marine life as part of an aquabiopolitical act of governance to serve human interests, wherein it is legitimate to kill or let some life die in favor of other (human) life. It is a politization of marine life.

Aquabiopolitics highlights, what Pauwelussen and Verschoor (2017) have explained, that shared worlds are not always terrestrial. That if we are to understand what happens in the marine world, we ought to start to think the terrestrial and marine world as closely interlinked (Strang 2023; Paerregaard and Uimonen 2021; Lien 2023). Thus, in this book, I use the term to bring together the anthropology of water, waste, and urban political ecology in showing how waste and pollutants move with relations of power from land to water and back. This movement includes other-than-human actors as well as different technological devices, and therefore theoretical inspiration is sought from STS and more-than-human studies to increase the understanding of how humans and other-than-humans create worlds mutually, but not equally. Moreover, as I stress the importance of visuality and invisibility, theoretical insights will be brought in from visual anthropology and media studies. This is important because scientists and scuba divers seek to make ongoing, but often hidden, aquabiopolitical relations visible in their quest to change the status quo. Thus, my approach is not firmly placed within *one* theoretical tradition, but is an eclectic collection of approaches that I view as necessary to make sense of the world that I have engaged in my research. An increasingly complex world begs for increasingly wide and sometimes almost contradictory perspectives to understand and explain it. This is my attempt to do so.

Book Outline

In this book, my aim is to open up the underwater world and make it visible to those living above the surface. The reader will sometimes join me below water, and at other times what is below will be brought up to the surface for scrutiny and engagement. I will show how human practices are decided by what is beneficial for humans, and that they govern on marine life to secure the well-being of humans.

Part I: The Scientists

In Chap. 2, "*Sediment Stories*," I introduce how I got access to the field of science. Then we join a group of marine scientists on a research expedition aboard one of Stockholm University's research vessels on the Baltic Sea, *Electra af Askö*. Then we follow the scientists and their work onboard, engaging with samples of sediment. After that, we follow the samples and the researchers back to their laboratories in the Netherlands, in Utrecht and Nijmegen, where further analysis and engagements occurs. In this chapter I seek to understand what relations, human and other-than-human, become visible when looking into the sediment below water. In Chap. 3, "*Technologically Sensing the Underwater World*," we return to the research vessel *Electra*, this time with another researcher and his students. In this chapter, the focus is on the technological affordances offered by the research vessel in order to open up the marine world. We will, as the scientist put it, "paint the ocean floor" and thereby provide a visual painting of the bottom for humans to engage. In Chap. 4, "*Slow Violence and the Plastisphere*," we examine one effect of aquabiopolitical relations: the rapidly developing marine issue of plastics. We look at the toxicity of plastics pollution for the marine environment, and how it has taken the shape of slow violence. I show how plastics can leak through waste management systems and end up in the marine environment, and continue with one initiative that uses the material capacities of jellyfish to construct a microplastics filter for wastewater treatment plants to capture plastics in the water. For this particular initiative, we travel to Trondheim, Norway, to visit the laboratories and meet with the scientists behind it. The first part ends with an interlude in Chap. 5.

Part II: The Trash Divers

In Chap. 6, "*No One's Water*," we examine the work of the trash scuba divers in Rena Mälaren. We join them below water and look into the practice of trash diving, then we see how relations between the divers, municipal workers, and politicians develop along with growing public support for the group. In Chap. 7, "*Looking into the Underwater World,*" Rena Mälaren stages a variety of interventions in their stand against the ongoing pollution of the water around Stockholm by the local authorities. Rena Mälaren addresses the long-time and ongoing harmful relationship between humans, waste, and water present in Stockholm while seeking to change the way that water is treated. These interventions take the shape of an underwater demonstration, waste sculptures and art. All of which is broadcast on social media as a way to engage the public in putting pressure on the authorities to act differently, to stop treating the water as a dump for unwanted things and substances. In Chap. 8, "*Trash Diving—a Global Comparison*," I conduct a comparative exercise that begins with the story of how Rena Mälaren seek to expand their collaborative relationships to include, for example, local fire departments, as well as the places where they are active. I then compare the work of Rena Mälaren with the work of trash scuba divers in Brazil and Thailand. Furthermore, the chapter illustrates how two international non-governmental organizations, Project AWARE and Trash Hero, are part of similar work as Rena Mälaren. The aim of the chapter is to highlight that waste in water is a global issue, as are the initiatives working against the increasing pollution of water around the globe. Furthermore, I emphasize how the materiality and sensorial experience of water differ between places, but that water is connected across the globe nevertheless.

Chapter 9: Ending

The final chapter revisits the initial notion of *how can the underwater world become visible, knowable, and relatable to humans?* It is a reflection on what we have learned about human politics and behavior from a perspective below the surface, focusing on the book's key research questions. The book ends with the most recent developments for the trash scuba divers with an example of how they turn from one type of waste removal to another, with work that brings them closer than ever to the scientific community and action against eutrophication and climate change.

References

Abasi, G., L. Li, and K. Breivik. 2019. Global Historical Stocks and Emissions of PBDEs. *Environmental Science & Technology* 53 (11): 6330–6340.

Agamben, G. 1998. *Homo Sacer: Sovereign Power and Bare Life*. Trans. Daniel Heller-Roazen. Stanford: Stanford University Press.

Alexander, C., and P. O'Hare. 2023. Waste and Its Disguises: Technologies of (Un)Knowing. *Ethnos. Journal of Anthropology* 88 (3): 419–443.

Anand, N. 2017. *Hydraulic City: Water and the Infrastructure of Citizenship in Mumbai*. Durham, NC: Duke University Press.

Aswani, S. 2020. New Directions in Maritime and Fisheries Anthropology. *American Anthropologist* 122 (3): 473–486.

Bakker, K.J. 2012. Water: Political, Biopolitical, Material. *Social Studies of Science* 42 (4): 616–623.

Ballestero, A. 2019a. The Anthropology of Water. *Annual Review of Anthropology* 48: 405–421.

———. 2019b. The Underground as Infrastructure? Water, Figure/Ground Reversals, and Dissolution in Sardinal. In *Infrastructure, Environment, and Life in the Anthropocene*, ed. K. Hetherington. Durham and London: Duke University Press.

Barnes, J. 2014. *Cultivating the Nile: Everyday Politics of Water in Egypt*. Durham NC: Duke University Press.

———. 2016. Uncertainty in the Signal: Modelling Egypt's Water Futures. *Journal of the Royal Anthropological Institute* 22 (S1): 46–66.

Barnes, J., and S. Alatout. 2012. Water worlds: Introduction to the special issue of Social Studies of Science. *Social Studies of Science* 42 (4): 483–488.

Björkman, L. 2015. *Pipe Politics, Contested Waters: Embedded Infrastructures of Millennial Mumbai*. Durham, NC: Duke University Press.

Blanchette, A. 2020. *Porkopolis. American Animality, Standardized Life, and the Factory Farm*. Durham: Duke University Press.

Briody, E., and T. Meerwarth Pester. 2014. The Coming of Age of Anthropological Practice and Ethics. *Journal of Business Anthropology* 1: 11–37.

Brown, K. 2020. *A Deep Divers Becoming*. BA Thesis: School of Sociological and Anthropological Studies: University of Ottawa.

Bubandt, N. 2023. Tidalectic Ethnography. Snorkeling the Coral Reefs of the Anthropocene. In *Rubber Boots Methods for the Anthropocene. Doing Fieldwork in Multispecies Worlds*, ed. N. Bubandt, A. Oberborbeck Andersen, and R. Cypher. Minneapolis: University of Minnesota Press.

Butler, J. 2010. *Frames of War. When Is Life Grievable?* London: Verso.

Chaline, E. 2017. *Strokes of Genius. A History of Swimming*. London: Reaktion Books.

Chao, S., K. Bolender, and E. Kirksey, eds. 2022. *The Promise of Multispecies Justice*. Durham: Duke University Press.

Chen, M. 2012. *Animaciest*. Durham: Duke University Press.

Collins, S., M. Durington, P. Favero, K. Harper, A. Kenner, and C. O'Donnell. 2017. Ethnographic Apps/Apps as Ethnography. *Anthropology Now* 9 (1): 102–118.

Cousteau, J. 1953. *The Silent World: A Story of Undersea Discovery and Adventure*. New York, NY: Harper & Brothers Publishers.

Davies, C. 2008. *Reflexive Ethnography: A Guide to Researching Selves and Others*. London: Routledge.

Dewan, C. 2020. Living with Toxic Development: Shipbreaking in the Industrializing Zone of Sitakunda, Bangladesh. *Anthropology Today* 36 (6): 9–12.

Dimmock, K. 2009. *Comfort in Adventure: the Role of Comfort, Constraints and Negotiation in Recreational SCUBA Diving*. PhD thesis. Lismore, NSW: Southern Cross University.

Dimmock, K., and E. Wilson. 2011. 'Take a Deep Breath': How Recreational Scuba Divers Negotiate In-Water Constraints. *Leisure/Loisir* 35 (3): 283–297.

Driessen, H., and W. Jansen. 2013. The Hard Work of Small Talk in Ethnographic Fieldwork. *Journal of Anthropological Research* 69 (2): 249–263.

Evans-Pritchard, E. 1940. *The Nuer: A Description of the Modes of Livelihood and Political Institutions of a Nilotic People*. Oxford: Clarendon Press.

Favero, P. 2007. 'What a Wonderful World!': On the 'Touristic Ways of Seeing,' the Knowledge and the Politics of the 'Culture Industries of Otherness'. *Tourist Studies* 7 (1): 51–81.

———. 2020. *Image-Making-India*. London: Routledge.

Ford, A. 2019. Embodied Ecologies. *Cultural Anthropology*, Fieldsights, 25 April. https://culanth.org/fieldsights/introduction-embodied-ecologies. Accessed 26 July 2023.

Fortun, K. 2001. *Advocacy after Bhopal. Environmentalism, Disaster, New Global Orders*. Chicago: University of Chicago Press.

Foucault, M. 1978. *The History of Sexuality. Vol. 1*. Trans. Robert Hurley. New York: Vintage.

Geismar, H., and H. Knox, eds. 2021. *Digital Anthropology*. 2nd ed. Abingdon, Oxon: Routledge.

Grasseni, C., ed. 2007. *Skilled Visions*. New York, Oxford: Berghahn.

Gupta, A., and J. Ferguson, eds. 1997. *Anthropological Locations. Boundaries and Grounds of a Field Science*. Berkeley, Los Angeles, London: University of California Press.

Hamdy, S.F. 2008. When the State and Your Kidneys Fail: Political Etiologies in an Egyptian Dialysis Ward. *American Ethnologist* 35 (4): 553–569.

Hammersley, M., and P. Atkinson. 2007. *Ethnography, Principles in Practice*. 3rd ed. London: Routledge.

Hannerz, U. 2004. *Foreign News: Exploring the World of Foreign Correspondents*. Chicago: University of Chicago Press.

———. 2006. Studying Down, Up, Sideways, Through, Backwards, Forwards, Away and at Home: Reflections on the Field Worries of an Expansive Discipline. In *Locating the Field: Space, Place and Context in Anthropology*, ed. S. Coleman and P. Collins. Oxford, New York: Berg.

———. 2010. *Anthropology's World: Life in a Twenty-First-Century Discipline*. London: Pluto Press.

Haraway, D. 2003. *The Companion Species Manifesto: Dogs, People, and Significant Otherness*. Chicago: Prickly Paradigm Press.

Harmanşah, Ö. 2020. Deep Time and Landscape History. How Can Historical Particularity Be Translated? In *Timescales Thinking Across Ecological Temporalities*, ed. B. Wiggin, C. Fornoff, and P. Kim. Minneapolis, London: University of Minnesota Press.

Hastrup, K., and F. Hastrup, eds. 2015. *Waterworlds: Anthropology in Fluid Environments*. Oxford: Berghahn.

Helmreich, S. 2009. *Alien Oceans. Anthropological Voyages in Microbial Seas*. Berkeley: University of California Press.

———. 2011. Nature/Culture/Seawater. *American Anthropologist* 113 (1): 132–144.

———. 2014. Homo Microbis: The Human Microbiome, Figural, Literal, Political. *Thresholds* 42: 52–59.

———. 2023. *A Book of Waves*. Durham: Duke University Press.

Hetherington, K. 2020. *The Government of Beans. Regulating Life in the Age of Monocrops*. Durham: Duke University Press.
Hjort, L., H. Horst, A. Galloway, and G. Bell, eds. 2017. *The Routledge Companion to Digital Ethnography*. New York: Taylor & Francis.
Holmbäck, G., and S.-O. Wallenstein. (trans.). 2014. *Michel Foucault. Biopolitikens födelse. Collège de France 1978–1979*. Hägersten: Tankekraft Förlag.
Hoover, E. 2017. *The River Is in Us. Fighting Toxics in a Mohawk Community*. Minneapolis, London: University of Minnesota Press.
Howe, C. 2019. *Ecologics. Wind and Power in the Anthropocene*. Durham: Duke University Press.
Ialenti, V. 2020. *Deep Time Reckoning. How Future Thinking Can Help Earth Now*. Cambridge, MA and London: The MIT Press.
Irvine, R. 2020. *An Anthropology of Deep Time. Geological Temporality and Social Life*. Cambridge: University of Cambridge Press.
Jue, M. 2020. *Wild Blue Media. Thinking Through Seawater*. Durham: Duke University Press.
Karlsson, B.G. 2015. Political Ecology: Anthropological Perspectives. In *The International Encyclopedia of the Social and Behavioural Sciences*, 2nd ed. Oxford: Elsevier Ltd.
Kirksey, E., and S. Helmreich. 2010. The Emergence of Multispecies Ethnography. *Cultural Anthropology* 25 (4): 545–576.
Knox, H. 2020. *Thinking Like a Climate. Governing a City in Times of Environmental Change*. Durham and London: Duke University Press.
Kohn, E. 2013. *How Forests Think. Toward an Anthropology beyond the Human*. Berkeley: University of California Press.
Krause, F. 2013. Seasons as Rhythms on the Kemi River in Finnish Lapland. *Ethnos. Journal of Anthropology* 78 (1): 23–46.
———. 2017. Towards an Amphibious Anthropology of Delta Life. *Human Ecology* 45 (3): 403–408.
Krause, F., and V. Strang. 2016. Thinking Relationships Through Water. *Society & Natural Resources* 29 (6): 633–638.
Larcher, J., and N. Oxley. 2015. Current Dilemmas of the Ethnographer behind the Camera. *Anthrovision* 3 (2).
Latour, B. 1988 [1984]. *The Pasteurization of France*. Trans. A. Sheridan and J. Law. Cambridge, MA: Harvard University Press.
Latour, B., and S. Woolgar. 1979. *Laboratory Life. The Construction of Scientific Facts*. Princeton, NJ: Princeton University Press.
Lemke, T. 2013. Foucault, Politics and Failure. In *Foucault, Biopolitics and Governmentality*, ed. J. Nilson and S.-O. Wallenstein. Huddinge: Södertörn University.
Lien, M. 2015. *Becoming Salmon. Aquaculture and the Domestication of a Fish*. Berkeley: University of California Press.

———. 2023. Fluid Scalability. *Ethnos. Journal of Anthropology* (online 11 Jul 2023).

Lien, M., and G. Pálsson. 2021. Ethnography Beyond the Human: The 'Other-than-Human' in Ethnographic Work. *Ethnos. Journal of Anthropology* 86 (1): 1–20.

Linton, J. 2010. *What Is Water? The History of a Modern Abstraction*. Vancouver: University of British Columbia Press.

Lokot, T. 2018. #IamNotAfraidToSayIt: Stories of Sexual Violence as Everyday Political Speech on Facebook. *Information, Communication & Society* 21 (6): 802–817.

Maaker, E. 2000. Integrating Ethnographic Research and Filmmaking: Video Elicitation for a Performance-oriented Analysis of the Teyyam Ritual. *Visual Anthropology* 12 (2): 186–197.

Marcus, G. 1995. Ethnography in/of the World System: The Emergence of Multi-sited Ethnography. *Annual Review of Anthropology* 24: 95–117.

Martin, E. 1996. Meeting Polemics with Irenics in the Science Wars. In *Science Wars*, ed. A. Ross. Durham and London: Duke University Press.

McCormack, F., and J. Forde. 2020. Fishing. In *Oxford Research Encyclopedia of Anthropology*. https://oxfordre.com/anthropology/view/10.1093/acrefore/9780190854584.001.0001/acrefore-9780190854584-e-183. Accessed 11 May 2021.

McDougall, D. 2005. *The Corporeal Image*. Princeton: Princeton University Press.

Merchant, S. 2011. The Body and the Senses: Visual Methods, Videography and the Submarine Sensorium. *Body and Society* 17 (1): 53–72.

Millar, K.M. 2018. *Reclaiming the Discarded. Life and Labor on Rio's Garbage Dump*. Durham: Duke University Press.

Minh, T., and N. Nguyen. 2018. *Waste and Wealth: An Ethnography of Labor, Value, and Morality in a Vietnamese Recycling Economy*. Oxford: Oxford University Press.

Montoya, T. 2017. Yellow Water: Rupture and Return One Year after the Gold King Mine Spill. *Anthropology Now* 9 (3): 91–115.

Nading, A.M. 2020. Living in a Toxic World. *Annual Review of Anthropology* 49: 209–224.

Nixon, R. 2011. *Slow Violence and the Environmentalism of the Poor*. Cambridge: Harvard University Press.

O'Hare, P. 2022. *Rubbish Belongs to the Poor: Hygienic Enclosure and the Waste Commons*. London: Pluto Press.

Oksala, J. 2013. Neoliberalism and Biopolitical Governmentality. In *Foucault, Biopolitics and Governmentality*, ed. J. Nilson and S.-O. Wallenstein. Huddinge: Södertörn University.

Olsen, B. 2021. The Char and the Oystercatcher. The Value of the Wild in Rossfjord, North Norway. *Ethnos. Journal of Anthropology*, Published Online May 09. https://doi.org/10.1080/00141844.2021.1924815.
Orlove, B., and S. Caton. 2010. Water Sustainability: Anthropological Approaches and Prospects. *Annual Review of Anthropology* 39: 401–415.
Paerregaard, K., and P. Uimonen. 2021. Water: An Anthropological Contribution. *Kritisk etnografi—Swedish. Journal of Anthropology* 4 (2): 9–13.
Paerregaard, K., A. Stensrud, and A. Andersen. 2016. Water Citizenship: Negotiating Water Rights and Contesting Water Culture in the Peruvian Andes. *Latin American Research Review* 51 (1): 198–217.
Paulson, S., and L. Gezon, eds. 2004. *Political Ecology across Spaces, Scales, and Social Groups*. London: Rutgers University Press.
Pauwelussen, A., and G. Verschoor. 2017. Amphibious Encounters: Coral and People in Conservation Outreach in Indonesia. *Engaging Science, Technology and Society* 3: 292–314.
Paxson, H. 2008. Post-Pasteurian Cultures: The Microbiopolitics of Raw-Milk Cheese in the United States. *Cultural Anthropology* 31 (1): 15–47.
———. 2012. *The Life of Cheese. Crafting Food and Value in America*. Oakland: University of California Press.
Pink, S. 2015. *Doing Sensory Ethnography*. London: SAGE Publications.
Pink, S., H. Horst, J. Postill, L. Hjorth, T. Lewis, and J. Tacchi. 2015. *Digital Ethnography. Principles and Practice*. London: SAGE Publications.
Podjed, D., and M. Gorup. 2021. Introduction: Why Does the World Need Anthropologists? In *Why the World Needs Anthropologists*, ed. D. Podjed, M. Gorup, P. Borecký, and C.G. Montero. New York: Routledge.
Puig de la Bellacasa, M. 2017. *Matters of Care. Speculative Ethics in More Than Human Worlds*. Minneapolis, London: University of Minnesota Press.
Raycraft, J. 2020. Seeing from Below: Scuba Diving and the Regressive Cyborg. *Anthropology and Humanism* 45 (2): 301–321.
Reed-Danahay, D. 2019. *Autoethnography. SAGE Research Methods*. London: SAGE Publications Ltd.
Renfrew, D. 2018. *Life Without Lead: Contamination, Crisis and Hope in Uruguay*. Berkeley: University of California Press.
Reno, J. 2015. Waste and Waste Management. *Annual Review of Anthropology* 44: 557–572.
———. 2016. *Waste Away. Working and Living with a North American Landfill*. Oakland, California: University of California Press.
Robins, P. 2020. *Political Ecology: A Critical Introduction*. 3rd ed. Oxford: Wiley.
Rodineliussen, R. 2017. *Divers Engaging Policy—Practices of Making Water*. Master's thesis. Stockholm University and Sida. http://urn.kb.se/resolve?urn=urn:nbn:se:su:diva-147293.

———. 2019a. [2017]. Visual Methods to Study the Underwater World: Scuba Divers and a Sensorial Experience of Water. *Anthrovision* 5(2).
———. 2019b. Ethnographic Exploration of a Sensorial Underwater World. *Irish Journal of Anthropology* 22 (1): 184–189.
———. 2019c. Organizing the Syrian Revolution—Student Activism through Facebook. *Visual Studies* 34 (3): 239–251. https://doi.org/10.1080/1472586X.2019.1653790.
———. 2021. Caring for Water: Underwater Waste, Trash Diving and Publicity in Stockholm. *Kritisk etnografi—Swedish. Journal of Anthropology* 4 (2): 67–81.
Rose, G. 2007. *Visual Methodologies: An Introduction to the Interpretation of Visual Materials*. London: Sage.
Rydén, F. 2014. Vårt Grisiga Hav. *Folke Rydén Productions*. https://vimeo.com/89600497. Accessed 26 July 2023.
Scaramelli, C. 2021. *How to Make a Wetland. Water and Moral Ecology in Turkey*. Stanford, CA: Stanford University Press.
Schmidt, J. 2019. *Water. Abundance, Scarcity, and Security in the Age of Humanity*. New York: New York University Press.
Schulthies, B. 2021. Phytocommunicability and Cross-Species Sociality. *Ethnos. Journal of Anthropology* 86 (2): 199–206.
Sennelart, M. (ed.). 2009. *Michel Foucault. Security, Territory, Population. Lectures at the Collège De France, 1977–1978*. Trans. G. Burchell. New York: Palgrave Macmillan.
Sharp, L. 2019. *Animal Ethos. The Morality of Human-Animal Encounters in Experimental Lab Science*. California: University of California Press.
Shurkin, N. 2009. *Animal Capital: Rendering Life in Biopolitical Times*. Minneapolis: University of Minnesota Press.
Strang, V. 2019. Water. In *The Cambridge Encyclopedia of Anthropology*, ed. F. Stein, S. Lazar, M. Candea, H. Diemberger, J. Robbins, A. Sanchez, and R. Stasch. https://doi.org/10.29164/19water.
———. 2021. Elemental Powers: Water Beings, Nature Worship, and Long-term Trajectories in Human-environmental Relations. *Kritisk etnografi—Swedish. Journal of Anthropology* 4 (2): 15–34.
———. 2023. Relating to the River: New Bio-cultural Diversities in Human Engagements with Water. *Kritisk etnografi—Swedish Journal of Anthropology* 5 (1): 33–47.
Thompson, M. 2017 [1979]. *Rubbish Theory. The Creation and Destruction of Value—New Edition*. London: Pluto Press.
Tousignant, N. 2018. *Edges of Exposure. Toxicology and the Problem of Capacity in Postcolonial Senegal*. Durham, London: Duke University Press.
Tsing, A. 2015. *The Mushroom at the End of the World: On the Possibility of Life in Capitalist Ruins*. Princeton, NJ: Princeton University Press.
Tsui, B. 2020. *Why We Swim*. Bungay: Clays Ltd.

van Ede, Y. 2009. Sensuous Anthropology: Sense and Sensibility and the Rehabilitation of Skill. *Anthropological Notebooks* 15 (2): 61–75.

van Helmond, N., B. Lougheed, A. Collebregt, F. Peterse, G. Fontorbe, D. Conley, and C. Slomp. 2020. Recovery from Multi-millennial Natural Coastal Hypoxia in the Stockholm Archipelago, Baltic Sea, Terminated by Modern Human Activity. *Limnology and Oceanography* 65: 3085–3097.

Vikström, K., and J. Wikner. 2019. Havets dolda andetag. *Havsutsikt*, nr. 2. https://www.havet.nu/havsutsikt/artikel/havets-dolda-andetag. Accessed 26 July 2023.

Wacquant, L. 2004. *Body and Soul*. Oxford: Oxford University Press.

Wakefield, S., and B. Braun. 2019. Oystertecture: Infrastructure, Profination, and the Sacred Figure of the Human. In *Infrastructure, Environment, and Life in the Anthropocene*, ed. K. Hetherington. Durham: Duke University Press.

Wallenstein, S.-O. 2013. Introduction: Foucault, Biopolitics, and Governmentality. In *Foucalt, Biopolitics and Governmentality*, ed. J. Nilson and S.-O. Wallenstein. Huddinge: Södertörn University.

Waltz, M. 2020. 'How It Is Grown Doesn't Matter, as Long as It's on the Table': Pesticide Use, Uncertainty and Future Aspirations. *Anthropology Today* 36 (6): 25–28.

Wanderer, E. 2020. *The Life of a Pest. An Ethnography of Biological Invasion in Mexico*. Oakland: University of California Press.

Weber, S. 2012. Bare Life and Life in General. *Grey Room* 46, Winter: 6–25.

Wenzel, Geissler P., and R. Prince. 2020. British Sofas and Global Toxic Flows. *Anthropology Today* 36 (6): 1–2.

Wolfe, M. 2017. *Watering the Revolution: An Environmental and Technological History of Agrarian Reform in Mexico*. Durham, NC: Duke University Press.

Wright, T. 2008. *Visual Impact: Culture and the Meaning of Images*. Oxford: Berg.

Wright, K. 2017. *Transdisciplinary Journeys in the Anthropocene. More-Than-Human Encounters*. New York: Routledge.

Wulff, H. 2008. To Know the Dancer: Formations of Fieldwork in the Ballet World. In *Knowing How to Know*, ed. N. Halstead, E. Hirsch, and J. Okely. Oxford: Berghahn.

PART I

The Scientists

CHAPTER 2

Sediment Stories

About halfway into the spring semester of 2019, I began to worry about gaining access to marine and plastic scientists in the field to which I aspired. I had emailed several scientists but received nothing even close to an affirmative reply. My fieldwork was set to begin in late summer 2019, so the outlook was dim. At this time, I was assigned to join a media workshop at Stockholm University that was to prepare researchers and staff to talk to the media. The workshop took place on campus on a rainy spring morning. I came a little early, so mostly the media crew and I were present. As the minutes passed, the room started to fill up with researchers from various disciplines: chemists, biologists, sociologists, psychologists, mathematicians, and others. I listened carefully for leads on what subjects these researchers worked with, hoping for someone from the marine sciences. As everyone sat around a large table, the media group introduced itself and asked the rest of us to do the same. I was first up and talked about my future research in a few words before the person to my right took over. I noticed that the chair to my left was empty and wondered if I should remove it. However, at this moment, a man entered, soaked from the rain, excused himself, and sat down in the chair next to me. He nodded in my direction in a silent greeting when he noticed I was looking at him. When the circle of introductions was about to conclude, only the man next to me remained. That was when my luck turned:

R. Rodineliussen, *Underwater Worlds*, Sustainable Development Goals Series, https://doi.org/10.1007/978-3-031-63370-6_2

> Hello everyone, my name is Thomas. I am the captain of Stockholm University's new research vessel *Electra*. As a captain, it is essential to know how to communicate with the media because I regularly will be questioned about the safety and other matters regarding the vessel. I also want to know more about how to best present my knowledge about research aboard the *Electra* to the general public. I am a great fan of science, but as I am no scientist, I understand the difficulties of translating scientific language into a language that the public, such as myself, can understand and appreciate.

I must have been staring because Thomas smiled in my direction and asked if everything was OK. I said that everything was more than OK and that I wanted to talk to him after the workshop. When we met afterwards, I told him about my project and my problems with gaining interest from scientists to collaborate. Thomas found this strange. As he explained to me, the Baltic Sea Center at Stockholm University has one of its stated new goals to start more collaborations with the social sciences. On this note, he promised to check with his team leader about upcoming expeditions on which he could invite me. Not long after, he arranged for me to join an expedition leaving in June 2019.

Meeting the Scientists

When June came, I took an early tram from the Stockholm city center to Djurgården, an island considered part of central Stockholm, where the research vessel *Electra* was docked. Thomas had told me to meet him, the mechanic, and the research crew of four members at 8 am. As usual, I was early, but the sun was shining, and I enjoyed a slow walk down the small road leading from the tram to the docks, where I found *Electra* waiting (Fig. 2.1). I went to her side and sat down on the quay, looking at her while listening to the waves, and tried to calm my nerves. I was a bit nervous; I really wanted this first meeting with the marine scientists to go well.[1] However, I told myself, why should it not? Thomas has already told them about me. If there were any objections, they should already have told him, right? Now I saw Thomas driving in, and he jumped out of the car, smiling and waving at me. "Great to see you Rasmus! This will be fun! But one thing: I know I promised you to inform the group about you.

[1] The scientists were from many different disciplines. On this first expedition, they came from biogeochemistry and microbiology. However, they referred to themselves as "marine scientists" to avoid having to specify their discipline too often. I will thus use this denomination when talking about them in general terms.

Fig. 2.1 The research vessel *Electra af Askö*. (Photo: Author)

Well, I forgot, but now you can do it yourself. And as a matter of fact, here they come!" … Gulp … if I was not nervous before, I really was now.

The group consisted of Niels van Helmond, Wytze Lenstra, Martijn Hermans, and Paula Dalcin Martins.[2] Thomas later explained that this was their second expedition on the *Electra*. They were, in fact, the first group to go out in her. After Thomas had greeted everyone, he pointed toward me and bluntly said: "This is Rasmus, a social scientist. He will join us today. Rasmus, please introduce yourself to the group while I get things going on the boat." I tried to tell the others about my project in as few

[2] The names of the scientists in this book are their real names. I have asked them if they wished to be anonymous. However, they have all asked to have their real names used as they view their participation as a form of outreach. Their work is also cited throughout this chapter.

words as possible hoping to capture their interest and not take too much time, as everyone was eager to get going. After my quick presentation, the group shook my hand and welcomed me on board. They told me they were OK with me joining them for the day. The expedition would continue for a week, but I figured today was a trial to see whether they could work with me around them or not.

I went to the upper deck to give the group space to unpack and to ask Thomas about the plans for the day and his role as captain. He explained that his current job with Stockholm University was the best one he ever had, although maybe not the most exciting locations (compared to some around Africa or polar expeditions). However, the *Electra* was never out to sea for more than 11 days at a time and had a more or less regular schedule, something he had never experienced before. Moreover, he enjoyed being the captain of the *Electra* because it was such a new and well-equipped vessel (the fact that he was involved in designing her might also have played a role in his admiration). "This makes being the infrastructure for research much more fun, especially because all the scientists are so happy about the experience!" *Electra* conducts research approximately 110 days a year, going to the northern reaches of the Baltic Sea and as far south as Kalmar and Simrishamn in southern Sweden. The rest of the time *Electra* functions as a transport vessel between Uttervik and the Baltic Sea Center's marine laboratory at Askö, an island in the Stockholm Archipelago.

When Thomas and I had finished our little chat, standing silently looking through the windows, one of the scientists, Paula, came up and started talking to me. She seemed interested in knowing more about my project and why I wanted to join their expedition. We spoke for a while and she told me about her work and what she hoped to accomplish on the voyage. Paula's work is related to microorganisms in ocean sediments that produce methane. They play a crucial role in climate change as methane is a highly potent greenhouse gas. Paula worries that due to increasing temperatures, methanogens might produce increasing amounts of methane. She wants to understand whether this is the case. If so, she will inform the public and authorities to hopefully help address the issue and change possible future scenarios. Paula aims to contribute to change that will help prevent or mitigate the effects of global climate change. Her efforts are visible in, among other things, a short information video she has published about her work (screenshot below). While Paula and I talked, we went out on the upper deck and stood there in the chilly morning breeze, looking down on the other three that ran around on deck to set up the different stations. Paula asked me how I planned on working and if I would primarily just observe. I told her that I would be delighted to help with whatever task they would entrust me with, explaining the basics of participant observation as a method. She laughed and said she would think about what I could help them with, and then went down to set up her gear. I remained where I was, allowing the group to settle in, while I rested my eyes on the shoreline of Djurgården that we slowly passed as *Electra* moved toward the archipelago.

A Brief Outline and Theoretical Note

Caterina Scaramelli (2021) has noted, along with Tanya Richardson (2018), that most theoretical anthropological work on water has focused on the fluid body (Ballestero 2019; Helmreich 2011; Stensrud 2016). However, Scaramelli suggests that the sediment on the bottom is just as important to take into account. This is because most "natural" and "unnatural" elements in water eventually end up in the sediment. Hence, to understand the development of water quality over an extended period of time, it is essential to look into the sediment. Scaramelli's argument stands out when looking at the practices of marine scientists and their focus on sediment core samples. Therefore, I follow Scaramelli and Richardson in this chapter, focusing on the materiality of the bottom when joining a group of marine scientists onboard the research vessel *Electra* during a week-long expedition to encounter sediment samples brought up from the seafloor. Then I will invite the reader to follow the samples from *Electra* to two research laboratories in the Netherlands, in Utrecht and Nijmegen. We will engage the sediment as an assemblage (Tsing 2015), where microbes and macro-organisms cohabit, live, and work.[3] Microbes can regulate the oxygen availability in the sediment, produce or consume methane, release or degrade toxins, and much more. The relations within the sediment and between the sediment and the surface world can be rendered visible when employing an aquabiopolitical lens that allows the mapping of various actors, human and other-than-human, and illustrates the cyclical nature and co-dependence inherent within these relations.

[3] I use the concept "assemblage" in accordance with Anna Tsing's (2015: 23–24) definition as a description of "open-ended gatherings," or "lifeways," wherein humans and other-than-humans constitute worlds mutually.

Furthermore, it becomes possible to see how other-than-humans work with, against, or for humans (Besky and Blanchette 2019), and provides an opportunity to learn about stories other than human ones (Rees 2019). Sediments can tell us a great deal about the human-waste-pollution-water relationship and the underwater world.

Cleaning the Cores: Ugh … Rotten Eggs!

After watching the work for a while, I walked around, asked a few questions, but it felt a little bit awkward. Not wanting to be annoying, but also not wanting to just be standing there looking over shoulders. I noticed the growing pile of dirty "cores" in the corner. Niels had cleaned a few earlier, but left some as he had to prepare for the next sampling station. Such cores can best be described as plastic tubes used to retrieve sediment from ~80 centimeters beneath the seafloor. I crossed the deck, took hold of the hose on the way, grabbed a core, and started cleaning. The rest of the sediment within the core did not move; even under the strong pressure of the hose; it was as if it was glued to the side of the core. Moreover, it had the most terrible smell; it reminded me of the "fun times" in the boys' locker room in school when there always was this one guy that had to fart out loud, screaming "rotten eggs" and laughingly running away. Well, that same scent came to me now. It is due to the high levels of sulfide in the sediment and water, as the group later explained to me. I scrubbed the first core clean, moving on to the next while getting wetter and wetter from the splashing when I failed to aim properly. With each splash of water, I also got more and more stains from the sediment on my clothes—the odor included. After about half an hour, Niels came up to me. "Thank you, Rasmus. This is helpful, as we really need those cores clean before we reach the next station, and we do not have enough staff to set up all gear, work with the samples, and clean at the same time. If you can help us with this, we will be super grateful!" So I continued cleaning, and when we reached the next station, Niels returned, "Rasmus, thanks! You want to help me take up the next couple of samples from the water?" In this way, I was allowed to assist more and more with taking up and handling the samples from the water.

Sampling with Cores

The *Electra* had a red machine on its right (starboard) side. Into this machine, two plastic cores with the length of one meter could be inserted.[4] The machine dropped the cores into the water, and with the help of the weight from the machine, the cores sank to the bottom and hit the sediment with enough force to penetrate it. When the cores were pulled up again, a lid slid down from the machine and covered the holes in the cores, ensuring that the sediment stayed intact and remained in place. These samples also contained about 20 centimeters of overlying water, or bottom water, as the water directly above the sediment is important for the understanding of findings within the sediment. Once the cores have resurfaced and are hanging above deck, the scientists take hold of one core and a rubber stopper that exactly fits the hole in the core. They use the stopper to pry the lid away and replace it with the stopper. Then the core can be removed from the machine and taken for analysis (Fig. 2.2).

The moment at which the stopper replaces the lid is crucial. All the sediment will pour out of the core and down onto the deck if it is done

Fig. 2.2 This screenshot from a video Paula did about her work shows a core going down from the boat toward the ocean floor. (Dalcin Martins 2019)

[4] Further reading on core sampling: van Helmond et al. (2020b: 2748).

incorrectly, requiring another core to be taken. Moreover, as it requires about 10 minutes to take a core sample in this area (there are examples from elsewhere where it takes hours, depending on the depth), the group cannot allow too many mistakes if they are to manage to sample all their stations within the time available to them.[5] After my initial work with cleaning the cores, Niels asked me if I would like to try to take a core out of the machine. Of course, I wanted to try! Thus, when the next pair of cores came up, I grabbed a stopper, and stood next to the machine, just as I had observed the others doing. Then I took hold of the core and used the stopper to remove the lid and place the stopper in its place. Or at least that was how I imagined it would play out! In reality, when the lid moved away, I heard a large swoop, and the next second, I was drenched in sediment to the laughter of the rest. Niels said, "Don't worry about it! All of us miss the cores now and then, especially in the beginning. Try another one. But be quicker and press harder next time." I repeated the procedure, took hold of the core; and this time did not wait for the lid to move. I just pressed the stopper in place, and it worked (Fig. 2.3).

Now I moved the core I managed to retrieve correctly to the "core holder." A core holder is a box with holes in the bottom and the top that holds cores stable so that they cannot fall, even if the boat rocks. Here Niels pointed to the uppermost layer of the core where I could see a thin but clearly distinguished brownish line. "It indicates that there is some oxygen and life in the bottom, here we can possibly find some macro-organisms." Then he pointed to another core next to it. "In this one the brownish layer does not exist, which means the bottom is not oxygenated." When looking at the cores, one could also observe bubbles moving from the bottom to the top and out of the sediment into the air. These were bubbles of methane gas that were being produced and released from the sediment.

Much can be learned about sediment just by looking at it, if one knows what to look for. Christina Grasseni (2004, 2007) suggests that the difference between the untrained and trained eye can be understood by the logic of "skilled visions." "Visions," importantly, is in the plural because it is not one type of vision but a context-dependent socialized vision. Grasseni (2007: 3) writes, "Cattle breeders, archaeologists, laser surgeons, even police consultants (Goodwin 1994) do each have a different world in

[5] In the next chapter, I will discuss the time frame further and the cost of renting the research vessel.

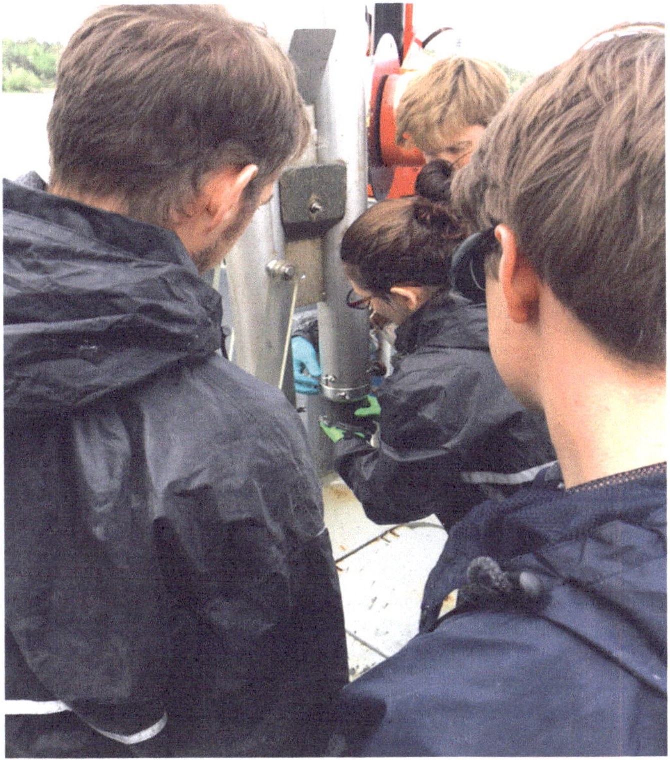

Fig. 2.3 Paula removes a core from the sampling machine while the group looks on. (Photo: Author)

front of their eyes, because they were each trained to see it differently." Further, Grasseni (2007: 4) notes, "*Skilled visions* are embedded in multisensory practices, where look is coordinated with skilled movement, with rapidly changing points of view, or with other senses, such as touch." Thus, embodying a skilled vision is not solely a practice of learning *to see*, but simultaneously it is a way *of seeing*. As Hayward (2010: 582) phrases it, "Sight is of the body, not just in the body." Niels and the others had developed specific skills in knowing how to look at and handle the cores. They combined their visual skills with other sensory factors such as smell to detect levels of sulfide in the bottom. By telling me what to look for, Niels was actively skilling my vision (Fig. 2.4).

Fig. 2.4 Cores in the "core holder." (Photo: Author)

After the cores were retrieved, it was time to start working with them. The different members of the group needed cores or portions of cores for different purposes. Some of the cores were placed in the large refrigerator on the boat. Others were taken for oxygen analysis. Some cores were "sliced" (I return to this below) to separate the layers from one another so that various tests could be run, depending on what the researchers were looking for in the different layers. For example, the top layer, down to a depth of 20 centimeters, was where macro-organisms could be found. That part would be put in a sieve and gently sprayed with water to remove the sediment and separate organisms like worms. They would be transferred in a solution called formaldehyde to die instantly and be preserved in that same state as they were found for later analysis and documentation. Paula described using formaldehyde as "we add what kills life."

Regarding the use of formaldehyde, Richard Irvine's (2020: 67) discussion on the lock-in of time is useful. Irvine describes a lock-in as an incapability to see and think beyond the present moment, using the example of a land preservation project. However, contrary to the case presented by Irvine, based on my ethnographic observations, I understand the idea of lock-in of time as a way to create the possibility of investigating the sediment further at another time. While, I agree with Irvine's characterization of pumping oil from the earth's crust as part of an "extraction from deep time" (Irvine 2020: 99), I view the "lock-in" as a way for the scientists to lock-in the present in order to look backward and forward in time (Ialenti 2020; Kress and Stine 2017; Wiggin et al. 2020). When the sampling machine takes a core from the bottom, it extracts a part of history (deep time) from the seafloor and brings it up to the surface for the scientists to interpret in the present. However, since many experiments can only be conducted in the laboratory, it is important to preserve samples for a later time when the needed equipment and tools will be available. This is the work of formaldehyde.

The macro-fauna station I assisted with was declared "Rasmus' Station." The sediment in the rinse shown on the left image below had a watery but gelatinous consistency. It was not easy to remove it through the tiny rinse holes while only using a gentle spray of water that could move the sediment around without risking hurting the organisms. At first, Paula showed me the procedure with a few samples before the task was handed over to me so that she could focus on helping the others. Most of Paula's work on the expedition was sampling and helping the others as her analysis and experimentation would occur later in the Netherlands. I found a large number of worms within the samples I rinsed, but also a lot of small bones (mainly from fish) and other remains of organic matter that had not yet fully degraded. Martijn told me that worms play a key role in regulating the availability of major elements in sediments. Worms are capable of creating burrows that restructure sediment and allow oxygen-rich water to reach deeper into the sea floor, a process known as bioturbation. The worms couple this with bioirrigation, where they ventilate their burrows with oxygen rich water. This facilitates the exchange of solutes between the sediment and the overlying water. In other words, worms can regulate the exchange of major elements between the sediment and overlying

Fig. 2.5 (Left) Sediment in the sieve at the macro-fauna station; Right: Worms in formaldehyde. (Photo: Author)

water. However, when large populations of worms are present, they can physically disturb the sediment, which in turn can negatively affect certain micro-organisms, such as cable bacteria, as will be discussed below (Fig. 2.5).

Core Slicing

One common method of studying sediment is core slicing. The sediment is pressed upward in the core so that a small part of it extends beyond the top of the core, depending on how large slices the scientist wants, usually about 2–3 centimeters. A plastic plate is then used to gently slice through the core, separating it from the rest. This part is moved to a centrifuge tube that separates the solid phase from the pore water, saving both for later analysis. The remaining part of the core is pressed up and subsequently cut. This continues until all of the sediment of the core is divided into small pieces. However, things do not always go as easy as planned. At this expedition, something was wrong with the push-up pole, the device used to press the sediment upwards. It could only do so halfway. Martijn went to the kitchen, collected a couple of soda cans, and started to place them between the pole and the stopper in the core. By this manual extending of the upward movement of the instrument, all of the core could be sliced even though the push-up pole did not cooperate as it was meant to. This idea by Martijn effectively solved the problem in one way, but the cans were nowhere near as stable a ground for the core to stand on as the pole. And as a result, the whole core wobbled around, and it would fall down the second no one held it. At this moment, I was the one standing holding the core. I ended up holding it still for another hour while it was sliced, then the next core, and the next after that. My shoulders were burning after a while as I had to keep my hands in a specific spot for such a long time.

From what I describe, it has become evident that the work of scientists in the field is both challenging and creative. This aspect is seldom brought out in scientific papers full of numbers, chemical symbols, and complicated names for methods. However, the data would not be collected without this hard work, creative solutions, and a bodily engagement with the materials scientists investigate. To be on the research vessel is a sensory experience (Pink 2015). Embodied knowledge and practices are paramount to learn about, and with, the scientists and their relation to the other-than-human world(s) (Ingold 2018; Helmreich 2009). When engaging the materiality of the sediment, by smelling, looking at, touching, and being touched by it, one comes to relate differently. Changing structures of the sediment tell different sediment stories. These stories relate to the human-waste-pollution-water relationship and hence become part of these relations. Varying levels of nutrients released from human societies will, for example, create less or more smell of sulfide—making the water more or less smelly, as a sign of "water quality" (Scaramelli 2013). This was pointed out to me by one of the researchers when *Electra* passed by a more affluent residential area:

> Here we know that the outflow of nutrients is high, and so are the levels of sulfide. In a few years the smell could be intense here. Making all these nice houses located in a place where it will smell like farts. But hey, it is their own waste they will feel.

Be Aware of the Oxygen!

Some samples taken from the cores must be kept away from oxygen as the sedimentary composition (e.g., minerals) will undergo chemical and physical changes upon contact with oxygen. Thus, these samples were placed in an air secure chamber. It was not so much a chamber as a balloon construction, basically a sealed plastic bag with two cavities for hands connected to a tube pushing gas into the bag on overpressure to make sure that no air could enter. The issue was that the bag had no exit valve. Eventually the bag became so full that it was in danger of exploding. At times, we could hear a loud and clear "Help!" from Niels, which meant that it was crucial that someone quickly came to open a minor hole in one of the corners of the bag to let some of the gas out, and then seal it before the whole procedure started over again. Meanwhile, Niels sorted and divided the samples into suitable tubes to bring back for analysis in the Netherlands (Fig. 2.6).

Another way of reducing the oxygen impact on the samples is to analyze them as quickly as possible. This was what Martijn did when testing

Fig. 2.6 Rear: Niels and his glove bag. Front: Paula marking samples. (Photo: Author)

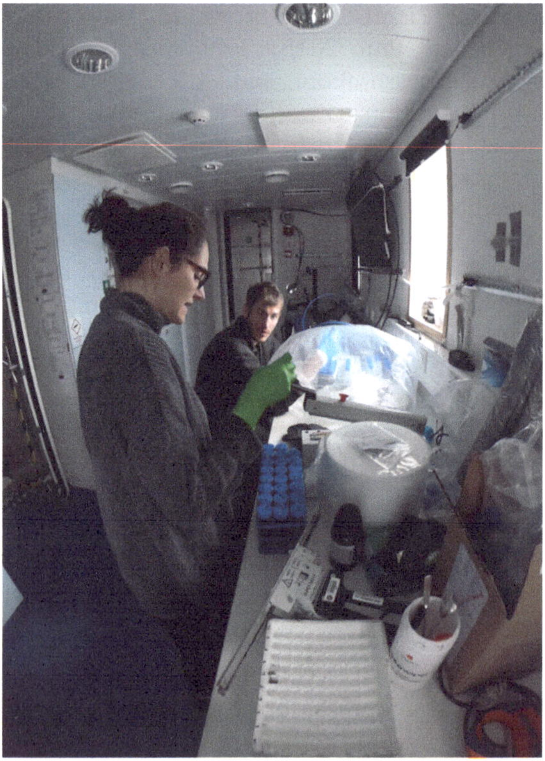

oxygen levels at the bottom with extremely thin sensors called micro-electrodes that were submerged into the sediment. He had to be very gentle when attaching them to the machine and placing them in the sediment samples within the core—something difficult to do on a moving ship. Martijn had to allocate areas of the bottom with oxygen in the sediment, as that makes it possible to find the cable bacteria that he studied. When I asked him to explain the role of cable bacteria in the sediment, he preferred to do so in writing (Fig. 2.7):

> Cable bacteria can have a profound effect on regulating the water quality in aquatic ecosystems. They are geographically widespread and have been found in freshwater, brackish, and marine sediments. Cable bacteria owe

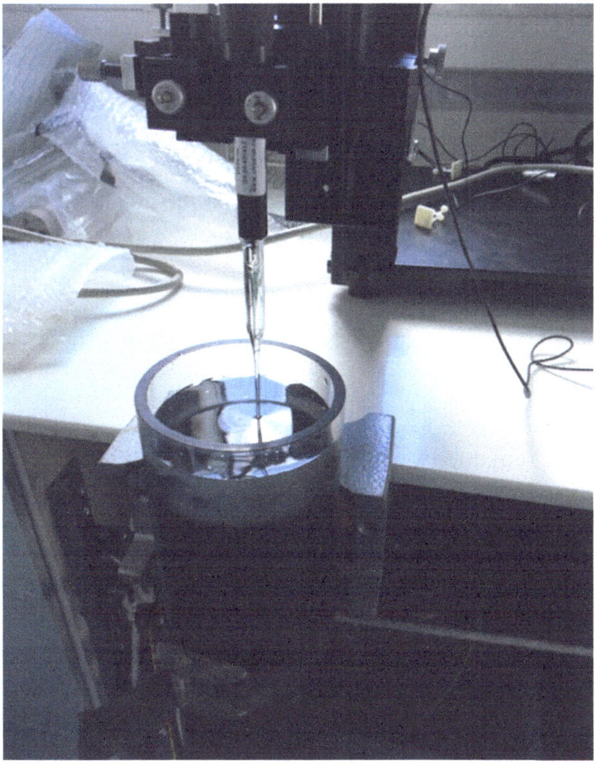

Fig. 2.7 A sensor taking an oxygen sample on *Electra*. (Photo: Author)

their name to the fact that they literally form bacterial wires in the sediment. These bacterial wires are highly conductive and are capable of transporting electrons in the sediment over cm-scale distances. Because of this electron transport, cable bacteria are capable of coupling the oxidation of sulfide to the reduction of oxygen. This results in a very strong pH gradient in the sediment. Near the sediment-water interface, the pH can reach values up to 9, whereas at a few cm depth it can go down to 6.5.

Typically, the strong pore water acidification at a few cm depth results in the dissolution of minerals, such as iron monosulfides, iron carbonates, and manganese carbonates. The dissolution of these minerals in turn leads to the release of dissolved iron and manganese in the pore water. When the dissolved iron and manganese diffuse upward, they can oxidize (i.e., conversion to iron oxides and manganese oxides) upon contact with oxygen from the overlying water. These iron oxides and manganese oxides can (temporarily) sequester both sulfide and phosphate, which implies that it will prevent the release of sulfide and phosphate from the seafloor into the water column. This is how cable bacteria can enhance the water quality and remediate eutrophication in the Baltic Sea.

There is no direct connection to climate change. However, global warming will result in stronger water column stratification, implying that there will be less mixing within the water column. Oxygen must be present in the bottom water (the water layer close to the seafloor) because cable bacteria need it for metabolism. In other words, global warming likely decreases the occurrence of cable bacteria.[6]

Daily Routine, Samples Out of the Sea, and What Then?

Throughout the week, we met daily at 7:30 am in the kitchen on *Electra*. The scientists remained on the ship overnight, while I went home in the evening to return early in the morning. During these meetings we discussed the past day and arranged the schedule for the day to come. Much of the planning often revolved around the weather because heavy rain would impact the sampling negatively—samples would be washed away. Once the stations for the day had been selected, the group started preparing their gear. On the first of these morning meetings, I was assigned the

[6] For more information on cable bacteria, see: Pfeffer et al. (2012); Meysman (2018); Kjeldsen et al. (2019). For dead zones: Hermans et al. (2019: 7494). On eutrophication: Hermans et al. (2019).

position of macrofauna person. Wytze said: "And Rasmus will continue with the filtering right? We will most likely have three new sediment samples for you today! Possibly, they will be oxygenated and might contain some worms!" From this, an outline of the oxygen levels in the Baltic Sea followed, and Niels took the lead in guiding me through three general scenarios: *Oxygenated* when there is oxygen. *Hypoxia* when there are low oxygen concentrations, and *Anoxic*, when there is no oxygen.[7] The last stage is a growing concern in the Baltic Sea. Niels continued to explain that excessive nutrients in the Baltic Sea are due to the human impact on the sea, primarily excess nutrients from agricultural runoff and wastewater treatment plants. These nutrients result in more algae in surface water that later sink to the bottom, where they are degraded by bacteria, consuming oxygen in the process, reducing oxygen levels and endangering the marine environment.[8]

When the expedition concluded, Niels and Wytze drove the samples back to the Netherlands, where they were stored for later analysis. In the autumn, I joined the researchers in their laboratories in Utrecht and Nijmegen to further analyze the samples. The rest of this chapter will be about this part of the journey.

Utrecht

The visit to Utrecht began with Martijn giving me a tour of the Utrecht University biogeochemistry laboratory. We passed climate controlled rooms that maintain the humidity and temperature at a constant level to facilitate incubation of samples during experiments. However, at the time of my visit they were also used as refrigerators to keep samples due to lack of storage space. Other rooms contained working spaces full of pipets and bottles, centrifuges, and a variety of machines. These included inductively coupled plasma-optimal emission spectroscopy (ICP-OES), inductively coupled plasma-mass spectroscopy (ICP-MS), and optical emission

[7] Further reading on Hypoxia: https://oceanservice.noaa.gov/hazards/hypoxia/ and Anoxic: https://bmcbiol.biomedcentral.com/articles/10.1186/1741-7007-8-30.

[8] For further reading on oxygen levels and nutrients from wastewater plants, see Hermans et al. (2019: 7495); van Helmond (2020a). For a historical overview of how human societies have impacted the Baltic Sea: Rydén (2014).

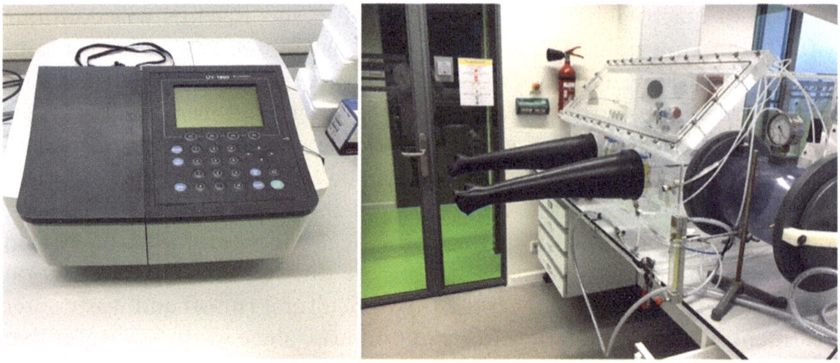

Fig. 2.8 Left: A Spectrometer; Right: A pressure chamber. (Photo: Author)

spectrometers of various quality and different operational costs.[9] As Niels would later say, pointing at the ICP-MS spectrometer, "We can only use every Euro once, so if we do not know exactly what we are looking for, we go for the cheaper machine first." The less expensive machine is also less precise, but can be engaged to narrow down a search before using the more accurate and expensive type to confirm the results. Martijn also showed me a spectrometer used to measure the concentration of a known substance in a solution, and a "glove box," or pressure chamber, similar to the one described earlier on the *Electra* (Fig. 2.8).

Meeting the Samples

After exploring the laboratories for a while, Martijn asked, "Shall we go meet the samples from *Electra*?" We exited the laboratory building and went to the storage room for core samples. The whole room held a cool temperature in order to keep the samples in good condition. There was also this particular smell of dried mud; it is rather subtle, but it lingers with you for a while even after exiting the room. The whole room had a unique

[9] "Mass spectrometry is an analytical tool useful for measuring the **mass-to-charge ratio** (*m/z*) of one or more molecules present in a sample. These measurements can often be used to calculate the **exact molecular weight** of the sample components as well. Typically, mass spectrometers can be used to identify unknown compounds via molecular weight determination, to quantify known compounds, and to determine structure and chemical properties of molecules." (https://www.broadinstitute.org/proteomics/what-mass-spectrometry)

feel as you walked past row after row of cores containing samples from sea beds from all over the world. It felt like the whole room was full of traces of history locked in time until analyzed in the laboratories. These samples will be molded into fine-grained powder and mixed with various chemical solutions to decipher their historical message.

In the vocabulary of art historian Ömür Harmanşah (2020), these cores could be seen as leaking deep time into the present. However, the scientists have the power to decide when and if these stories will be heard (for another example of how time and knowledge is controlled, see Telesca 2020). This power relation in science was noted well before by Steve Fuller in the volume *Science Wars*, when he asked, "Is Science special because of how much it can explain or how much it can control?" (1996: 31). According to Fuller, the role of the scientist as the interpreter of natural phenomena has increasingly become that of predicting future scenarios. In the same volume, Hilary Rose (1996: 85) notes that greater public literacy in scientific matters has led to growing public skepticism about science. If thinking with Emily Martin (1996: 64) this leads to the question of, "What counts as knowledge and who gets to determine this?" Vincent Ialenti (2020) notes similar issues as Fuller, Rose, and Martin when he talks about a crisis of trust for scientific expert knowledge among the public, as their own and scientists' predictions not always follow one another. Another concern regards whether scientists really tell the whole story, or if they are controlling the story as per Fuller's question. This is worth thinking about when one contemplates the storage room full of unanalyzed cores. We shall return to this in Chap. 3 (Fig. 2.9).

We walked by row after row of stored cores. Many were far larger than the ones we had collected on *Electra*. Some, up to 20 meters long, were divided into 1-meter segments. On most expeditions cores are taken by each research group at every sampling location (also called station). However, there are situations where the sampling procedure takes too long to procure one for each group. In that case, sub-samples from the sliced core may be distributed to different teams. Thus, each research group may get duplicate samples from the same depth cores. Further into the room, Martijn stopped and pointed at one row of standing cores and a white box in the middle of the corridor, "Those are the samples from Electra!" (Fig. 2.10).

Fig. 2.9 Shelves full of core samples. (Photo: Author)

Experiments in the Laboratory

After my initial meeting with the *Electra* samples, I joined Niels for another tour of the laboratories. A master's student in the laboratory was occupied with removing small plastic pipes containing sediment and liquid from a centrifuge. The first step in processing core samples is to separate the pore water from the sediment by centrifugation. The next step is that the remaining liquid in the sediment samples is removed by freeze drying. Later, the sediment samples are pulverized using a mortar and pestle inside a glove box. After that, different phosphorus fractions are extracted sequentially by using a variety of solvents. The most difficult fraction to extract is organic phosphorus because some of it is located within organic complexes. In order to destroy the organic complexes, the samples are reduced to ashes in an oven, often at 550 degrees Celsius for several hours (Fig. 2.11).

Fig. 2.10 Left: Standing core samples from *Electra;* Right: Box with the sliced samples from the same expedition. (Photo: Author)

When the sediment is ashed correctly, it is possible to remove the last of the phosphorus and thus obtain a complete picture of how much of it and in what form the sea has been able to bury at that location. The water left after separation from the sediment is then run through a spectrometer (shown above), to search for additional phosphorus. It is essential to know how much phosphorus can be buried in the sediment because scientists can thereby calculate the effects of eutrophication.

Niels then showed me a device in which sediment is freeze dried. As ice begins to form, it is removed by a vacuum pump without the ice turning into a liquid. This is crucial for many of the experiments as the samples must be in the form of a dry powder to mix them with various types of acids and solutions. After the sediment is freeze dried, it is placed in a glove box without any contact with oxygen.[10] The sample is first placed in a sluice on the right side of the box. When the samples are inside the sluice, the whole thing is rinsed with nitrogen gas while oxygen is sucked

[10] Further reading: Hermans et al. (2019: 7498).

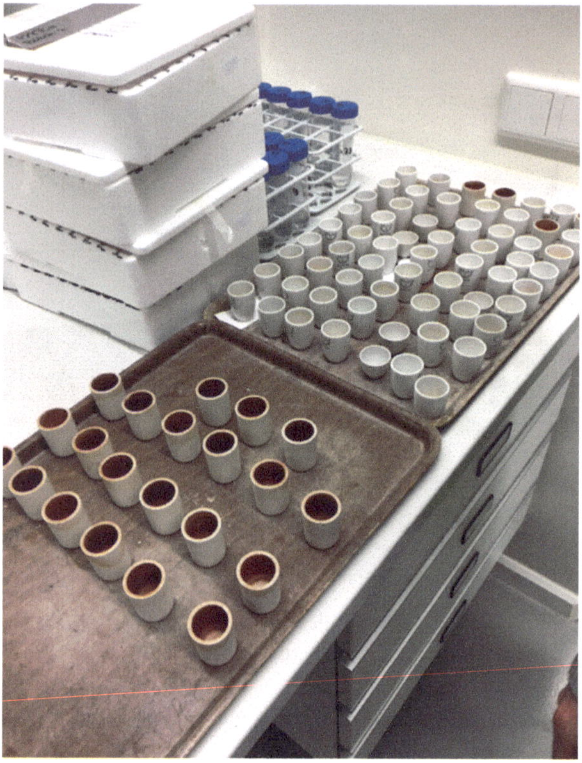

Fig. 2.11 Ceramic crucibles that hold sediment during ashing. (Photo: Author)

out. This is repeated three to four times until they are sure no oxygen remains. When all oxygen is gone, the door to the sluice is opened, and the samples are brought into the glove box interior before the door to the sluice is closed again. As a last precaution, before the samples are opened, the level of oxygen in the box is measured. If it is a higher concentration than 0.3%, the whole box is again flushed with nitrogen. When the oxygen level in the box is zero, the samples are opened and grounded to make certain they are in the form of a fine powder, which is then poured into numbered pipes and sealed. These will be stored inside the glove box until they are needed for an experiment (Fig. 2.12).

Niels also showed me a collection of bottles filled with sediment and bottom water. "Do you remember these from *Electra*? Me and Wytze

Fig. 2.12 Left: Freeze dried samples; Right: Niels working in the glove box. (Photo: Author)

took samples from this core with tape and holes, from the different sediment layers." These samples were used to measure levels of methane at various depths in the sediment.[11] Niels turned one of the bottles over and showed what he called a "head space," needed to enable taking small quantities of a sample out of the bottle. A needle is used to insert nitrogen and withdraw water. This is placed in an instrument called a trace gas chromatograph, which is equipped with a flame ionization detector. It calculates the concentrations of gases, such as methane. Based on the age of the sediment, it is then possible to understand how much methane this particular sediment produces, allowing the scientists to make assumptions about future scenarios and the role of the ocean as a source of methane (Fig. 2.13).

My visit to Utrecht revealed the importance of advanced technology in the scientists' interaction with the sediments. They use various technical equipment to store and analyze their material and trust the information that these provide about sediment age, methane production, phosphorus sequestration and burial, oxygen consumption, and sulfide production.

[11] Further reading: van Helmond et al. (2020b: 2749); Lenstra et al. (2018).

Fig. 2.13 One of the bottles described above. (Photo: Author)

They are also using the uncertainty of some of the technology (the cheaper Optical Emission Spectrometer, for example) to narrow down their research span so that they can make more well-educated guesses to test with more precise instruments and experiments. Cymene Howe (2019: 74) has elaborated on the role of machines, or more precisely trucks, in wind-power management. She defines trucks as "machinic devices that establish and sustain relationships" between people and the environment. The advanced technical equipment in the lab and on *Electra* has similar qualities, without which this kind of research would be impossible. The scientists put trust in the detailed results from some instruments or use the wider and less precise results of other instruments to gain an advantage. Knowledge is thus technologically mediated in the laboratory, emphasizing the importance of the role of technology in the present study.

This technologically mediated knowledge generated in the laboratory warrants to repeat Emily Martin's (1996: 62–62) questioning of where and how knowledge is produced. She argues that feminist scholars have "changed what counts as the production of knowledge by transforming (not just adding to) the very tenet of what counts as knowledge." Following this line, it is paramount to emphasize that these instruments which I discuss are active agents that contribute to the production of knowledge (Candea 2016), as illustrated by Latour and Woolgar in their book *Laboratory Life* (1979). These instruments are used by scientists to make visible a "truth" presumably already present "out there" in nature. This truth is then communicated to other scholars, governments, and the public as scientific facts. However, as I have shown, and will elaborate further, this process of knowledge-making in the laboratory is not unilineal; it requires that we accept the idea that there is a truth out there waiting to be discovered. For, as Andrew Ross notes,

> Such studies demonstrated that scientific knowledge is not given by the natural world but is produced or constructed through social interactions between/among scientists and their instruments, and that these interactions are mediated by the conceptual apparatuses created in order to frame and interpret the results. (1996: 12)

Scientists must make decisions on where to take samples, which samples to run through their instruments, and when to do this. There is a margin of error in the information generated by the instruments, and there are other possible factors that can alter the end results. Experiments are constructed to allow repetition that will cross-check results (according to positivist principles). However, there are also other elements involved. The choices scientists make are what effectively causes science to be political, as different decisions can produce alternate versions of the world and render different beings visible and knowable (Candea 2016; Liboiron 2021).

Nijmegen

I also visited Paula in the microbiology department of Radboud University in Nijmegen. When I asked her to reiterate the aim and topic of her research, she told me that methane is the second most important greenhouse gas with regard to effects on climate change. Amounts of methane have dramatically increased over the last 100 years, attracting scientists' interest in understanding why this is so. She explained that during the expedition on *Electra*, the goal was to take samples in areas with different impacts from nutrients and varying levels of oxygen in order to investigate how deoxygenation and eutrophication affects methane production. The results indicated that increasing eutrophication and lower levels of oxygen in bottom water do in fact increase the production of methane. This is a process that then causes a so-called feedback loop: methane heats the atmosphere, triggering further heating, increasing eutrophication and reducing amounts of oxygen in the water, which again leads to further methane production.

Paula seeks to understand this process by close engagement with methanogens in the bottom sediment. In what follows, we accompany Paula into the microbiological laboratory, to learn more about her work's practical and relational aspects. We started in a common workroom where she showed me the basic tools, such as bottles, clinically clean water, pipettes, centrifuges, shaking tables, and scales. Researchers brought in their samples and the chemicals they needed to do work. In the next room, we found a gas table with taps for all types of gases that might be needed, also a temperature-controlled shaking table, and glove boxes (Fig. 2.14).

We then entered a room with living cultures of microorganisms. To the right, Paula pointed toward a cultivation of anammox (anaerobic ammonium oxidation). Anammox put the microbiology department at Radboud University on the global map as it was here that it was first produced. It is a bacterial culture used in wastewater treatment plants that mimics the natural nitrogen cycle of the ocean. It helps to minimize the anthropogenic influx of nutrients into the sea and thus lessens the impact of eutrophication.[12] Not

Fig. 2.14 Left: Open lab space; Center: Temperature-controlled shaking table; Right: Gas table. (Photo: Author)

[12] Further reading on annamox, see Kuenen (2008).

far from the anammox reactor, Paula showed me what she called "a very expensive machine." It can keep samples of living bacteria and organisms under the same conditions that they would experience if they were still in the Baltic Sea. The machine allows Paula to keep a cultivation for observation in the same conditions as from where they were sampled. This machine is one of the key reasons that Paula chose this laboratory. The cultivation that Paula worked on was very slow-growing, needing years to become a usable quantity. This meant that she had to be very careful with it. However, experiments involve risks. As Paula noted, "I am actually doing an experiment right now that could kill the cultivation if I am very unlucky. But I have reserved a part of it to have in case all goes badly." (Fig. 2.15).

This risk of mortality of other-than-humans in experiments is something that, for example, Lesley Sharp has written about in her book *Animal Ethos* (2019). She illustrates the relationship of "care" between scientists and laboratory animals, one that inevitably ends with the animal's death, no matter how well cared for it is. Sharp (2019: 157) writes, "Among the more common challenges associated with working with animals is their ability to inspire emotional responses in their human cohabitants." Although I would describe Paula's relationship with her microorganisms as one of care and compassion, it might be stretching this relationship to consider it akin to a human-animal relationship. Thus, although we can see a relational bond at play, different levels of affection should be noted, especially if thinking in terms of good or bad care, that is, what form care should take and who

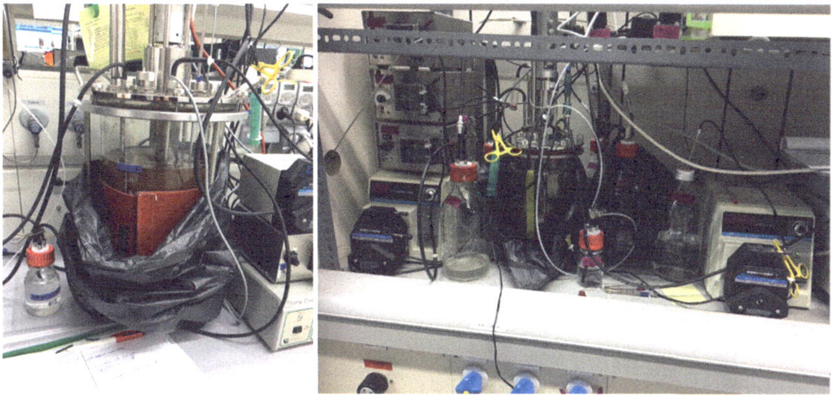

Fig. 2.15 Left: Anammox culture generator; Right: The Baltic Sea conditions generator. (Photo: Author)

should be the recipient and provider (McKearney and Amrith 2021). This discussion will be continued in Chap. 4.

When I asked Paula about her relationship with her microorganisms, she called them her Tamagotchi (digital pet). She drew this parallel between her organisms and that old toy because "they have very similar needs. Every now and then I have to check in on them to see that they have enough food and that they are feeling well, and give them some attention. Just like it was when I played with my Tamagotchi when I was little!" Checking in on her organisms is a type of care (Sharp 2019: 19). This practice entailed taking a small sample from their container and applying it to a pH stick to measure the pH level (Fig. 2.16). Although the machine runs on the same settings all the time, the pH somehow changes,

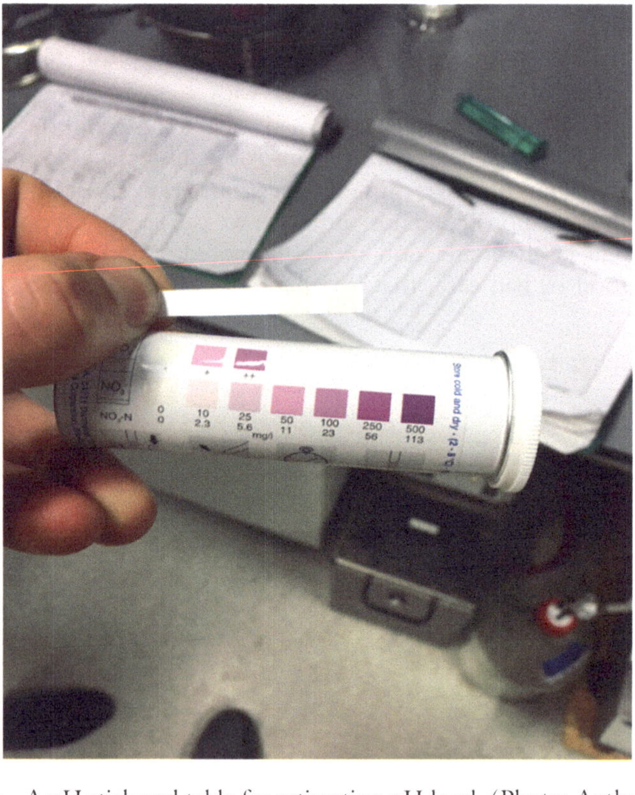

Fig. 2.16 A pH stick and table for estimating pH level. (Photo: Author)

Fig. 2.17 Paula and I among the living cultures. (Photo: Paula)

and if it does, the organisms risk dying. Therefore, at least once a day, preferably more often, their pH level needs to be checked and the machine adjusted. When Paula, for example, goes on vacation or on an expedition, someone else will need to do this for her, which means that the people in the group have to cover for one another. The fact that the people in the group have to cover for one another emphasizes the need to maintain a collegial working atmosphere. The researchers care for their microorganisms jointly (Fig. 2.17).

Forcing a Human Reaction?

Paula told me that the samples retrieved from *Electra* are currently under active observation as they have produced large amounts of methane since arriving to the laboratory. The question is how long they can continue producing methane outside of their normal environment. A prolonged capacity to do so would confirm strong methane-producing power. This indicates that the amounts of methane in the waters around Stockholm are high. According to Paula, it is something in the relationship between the environment and the humans around Stockholm that is clearly producing a lot of methane. It is worrying because it highlights previous and ongoing practices within human societies that cause the methanogens to be more productive, and this can result in escalating climate change. As Paula and her colleagues write in one article based on the fieldwork we did together,

> Coastal zones account for significant global marine methane emissions to the atmosphere. In coastal ecosystems, the tight balance between microbial methane production and oxidation in sediments prevents most methane from escaping to the water column. Anthropogenic activities, causing eutrophication and bottom water deoxygenation, could disrupt this balance in the microbial methane cycle and lead to increased methane release from coastal sediments. Here, we combined microbiological and biogeochemical analyses of sediments from three sites along a bottom water redox gradient (oxic-hypoxic-euxinic) in the eutrophic Stockholm Archipelago to investigate the impact of anthropogenically-induced redox shifts on microbial methane cycling. (Dalcin Martins et al. 2022)

If we follow Bradshaw's (2022: 291) suggestion and listen to microorganisms on their own terms, and respond "to the agency of these organisms, responding to their responses to a polluted earth," then we might realize that our current treatment of the Baltic Sea is far from sustainable—creating the situation described by Paula and her colleagues. Natasha Myers (2019: 127) suggests that the "management of plants in a garden does set in motion particular forms of labor and care, and particular forms of governance that dictate who and what lives and dies within its enclosures …. Each garden thus enacts its own form of biopolitics." This is biopolitics, as in aquabiopolitics, where the focus is on the processes of human-water relations enacted within the confines of the Baltic Sea, where life is regulated in a way that focuses on human survival. The currently enhanced methane-producing power of the sediments around Stockholm can provide an incentive for governmental regulations to govern over life in the water to secure human life on land.

What about other-than-human life? That methanogens produce such amounts of methane is a response to the way water is treated and a sign of a distressed marine world. It is common knowledge that the Baltic Sea is extremely polluted, and yet, cruiser ships continue to dump black water (poo water) despite bans to do so, countries allow unclean water to pour out into the sea via rivers and waterways, and keep on fishing beyond the limits of fish stock. If we were to actually listen to, and respond with these methanogens, we would need to change all this, and more. And it is becoming urgent: studies show that eutrophication will increase methane production and its release into the atmosphere by approximately 30–90% over the next decade, equivalent to 18–33% of annual carbon dioxide emissions from fossil fuels. Thus, limiting eutrophication is not only good for the quality of and life within water, but also important in fighting escalating climate change (Beaulieu et al. 2019).

The relationship between Stockholm and the Baltic Sea is a clear example of what I describe as aquabiopolitical relations. For as Paula and colleagues show, these are relations that effectively decrease the level of life in the Baltic Sea, and have done so for a long time, often with deadly consequences for numerous types of life. Take cod, halibut, and eel as examples of fish that are under threat of disappearing from Swedish waters, mainly due to overfishing, but also as a result of decreasing oxygen levels following increasing eutrophication. When effects have started to rebound to human societies in the form of increasing release of methane from the water, less food for consumption, or increasing smelliness of water, it results in humans seeking to change these relationships (mainly through policy regulations). This follows what I outline in the introductory Chap. 1. "*Diving in*" namely that marine life becomes important for humans if it affects human life. For example, the way methanogens act does now indeed affect human life—making methanogens into something more than a bare life, as something humans should care about. When looking at these relationships, it becomes evident that it is human actions and decisions that shape the situation, but that some other-than-humans can react in ways not always aligned with human wishes, and thereby force some change as to how humans treat water and the beings within. Sediments truly have many stories to tell, and humans would do well to listen.

References

Ballestero, A. 2019. The Anthropology of Water. *Annual Review of Anthropology* 48: 405–421.

Beaulieu, J., T. DelSontro, and J. Downing. 2019. Eutrophication Will Increase Methane Emissions from Lakes and Impoundments During The 21st Century. *Nature Communications* 10: 1375.

Besky, S., and A. Blanchette, eds. 2019. *How Nature Works. Rethinking Labour on a Troubled Planet.* Albuquerque: University of New Mexico Press.

Bradshaw, A. 2022. Can Microbes Be Active Participants in Research? Developing a Methodology for Collaborating with Plastic-Eating Microbes. *Environmental Humanities* 14 (2): 284–302.

Candea, M. 2016. Science. *The Open Encyclopedia of Anthropology*. https://www.anthroencyclopedia.com/entry/science.

Dalcin Martins. 2019. https://www.youtube.com/watch?v=kxQ1l__Lr3U. Accessed latest 22 July 2024.

Dalcin Martins, P., J. de Monlevad, W. Lenstra, A. Wallenius, M. Echeveste Medrano, M. Hermans, C. Slomp, C. Welte, M. Jetten, and N. van Helmond. 2022. Sulfide Toxicity as Key Control on Anaerobic Oxidation of Methane in Eutrophic Coastal Sediments. *BioRxiv*. https://doi.org/10.1101/2022.02.10.479873.

Fuller, S. 1996. Does Science Put an End to History, or History to Science? Or, Why Being Pro-science Is Harder than You Think. In *Science Wars*, ed. A. Ross. Durham and London: Duke University Press.

Goodwin, C. 1994. Professional Vision. *American Anthropologist* 96 (3): 606–633.

Grasseni, C. 2004. Skilled Vision: An Apprenticeship in Breeding Aesthetics. *Social Anthropology* 12 (1): 41–55.

———., ed. 2007. *Skilled Visions*. New York, Oxford: Berghahn.

Harmanşah, Ö. 2020. Deep Time and Landscape History. How Can Historical Particularity Be Translated? In *Timescales Thinking Across Ecological Temporalities*, ed. B. Wiggin, C. Fornoff, and P. Kim. Minneapolis, London: University of Minnesota Press.

Hayward, E. 2010. Fingeryeyes: Impressions of Cup Corals. *Cultural Anthropology* 25 (4): 577–599.

Helmreich, S. 2009. *Alien Oceans: Anthropological Voyages in Microbial Seas*. University of California Press.

———. 2011. Nature/Culture/Seawater. *American Anthropologist* 113 (1): 132–144.

Hermans, M., W. Lenstra, S. Hidalgo-Martinez, N. van Helmond, R. Witbaard, F. Meysman, S. Gonzalez, and C. Slomp. 2019. Abundance and Biogeochemical Impact of Cable Bacteria in Baltic Sea Sediments. *Environmental Science & Technology* 53: 7494–7503.

Howe, C. 2019. *Ecologics. Wind and Power in the Anthropocene*. Durham: Duke University Press.

Ialenti, V. 2020. *Deep Time Reckoning. How Future Thinking Can Help Earth Now*. Cambridge, MA and London: The MIT Press.

Ingold, T. 2018. *Anthropology: Why It Matters*. Polity.

Irvine, R. 2020. *An Anthropology of Deep Time. Geological Temporality and Social Life*. Cambridge: University of Cambridge Press.

Kjeldsen, K., L. Schreiber, C. Thorup, T. Boesen, J. Bjerg, T. Yang, M. Dueholm, S. Larsen, N. Risgaard-Petersen, M. Nierychlo, M. Schmid, A. Bøggild, J. van de Vossenberg, J. Geelhoed, F. Meysman, M. Wagner, P. Nielsen, L. Nielsen, and A. Schramm. 2019. On the Evolution and Physiology of Cable Bacteria. *PNAS* 116 (38): 19116–19125.

Kress, W., and J. Stine. 2017. *Living in the Anthropocene: Earth in the Age of Humans.* Washington, DC: Smithsonian Books.

Kuenen, J. 2008. Anammox Bacteria: From Discovery to Application. *Nature Reviews Microbiology* 6: 320–326.

Latour, B., and S. Woolgar. 1979. *Laboratory Life. The Construction of Scientific Facts.* Princeton, NJ: Princeton University Press.

Lenstra, W., M. Egger, N. van Helmond, E. Kritzberg, D. Conley, and C. Slomp. 2018. Large Variations in Iron Input to an Oligotrophic Baltic Sea Estuary: Impact on Sedimentary Phosphorus Burial. *Biogeosciences* 15: 6979–6996.

Liboiron, M. 2021. *Pollution Is Colonialism.* Durham and London: Duke University Press.

Martin, E. 1996. Meeting Polemics with Irenics in the Science Wars. In *Science Wars*, ed. A. Ross. Durham and London: Duke University Press.

McKearney, P., and M. Amrith. 2021. Care. *The Open Encyclopedia of Anthropology.* https://www.anthroencyclopedia.com/entry/care. Accessed 26 July 2023.

Meysman, F. 2018. Cable Bacteria Take a New Breath Using Long-Distance Electricity. *Trends in Microbiology* 26 (5): 411–422.

Myers, N. 2019. Five. From Edenic Apocalypse to Gardens against Eden: Plants and People in and after the Anthropocene. In *Infrastructure, Environment, and Life in the Anthropocene*, ed. K. Hetherington. Durham: Duke University Press.

Pfeffer, C., S. Larsen, J. Song, M. Dong, F. Besenbacher, R.L. Meyer, K. Kjeldsen, L. Schreiber, A. Gorby, M. El-Naggar, K. Leung, A. Schramm, N. Risgaard-Petersen, and L. Nielsen. 2012. Filamentous Bacteria Transport Electrons over Centimetre Distances. *Nature* 491: 218–221.

Pink, S. 2015. *Doing Sensory Ethnography. 2nd Ed.* SAGE.

Rees, T. 2019. For Microbes. Theorizing the Contemporary. *Cultural Anthropology. Fieldsights*, April 25. https://culanth.org/fieldsights/for-microbes. Accessed 26 July 2023.

Richardson, T. 2018. The Terrestrialization of Amphibious Life in a Danube Delta 'Town on Water.' *Suomen Antropologi: Journal of the Finnish Anthropological Society* 43(2): 3–29.

Rose, H. 1996. My Enemy's Enemy Is—Only Perhaps—My Friend. In *Science Wars*, ed. A. Ross. Durham and London: Duke University Press.

Ross, A., ed. 1996. *Science Wars.* Durham and London: Duke University Press.

Rydén, F. 2014. Vårt Grisiga Hav. *Folke Rydén Productions.* https://vimeo.com/89600497. Accessed 26 July 2023.

Scaramelli, C. 2013. Making Sense of Water Quality: Multispecies Encounters on the Mystic River. *Worldviews* 17: 150–160.

———. 2021. *How to Make a Wetland. Water and Moral Ecology in Turkey.* Stanford, CA: Stanford University Press.

Sharp, L. 2019. *Animal Ethos. The Morality of Human-Animal Encounters in Experimental Lab Science*. Oakland, CA: University of California Press.

Stensrud, A. 2016. Climate Change, Water Practices, and Relational Worlds in the Andes. *Ethnos. Journal of Anthropology* 81 (1): 75–98.

Telesca, J. 2020. Fishing for the Anthropocene. In *Timescales Thinking Across Ecological Temporalities*, ed. B. Wiggin, C. Fornoff, and P.E. Kim. Minneapolis, London: University of Minnesota Press.

Tsing, A. 2015. *The Mushroom at the End of the World: On the Possibility of Life in Capitalist Ruins*. Princeton, NJ: Princeton University Press.

van Helmond, N., B. Lougheed, A. Collebregt, F. Peterse, G. Fontorbe, D. Conley, and C. Slomp. 2020a. Recovery from Multi-millennial Natural Coastal Hypoxia in the Stockholm Archipelago, Baltic Sea, Terminated by Modern Human Activity. *Limnology and Oceanography* 65: 3085–3097.

van Helmond, N.A.G.M., E.K. Robertson, D.J. Conley, M. Hermans, C. Humborg, J. Kubeneck, W.K. Lenstra, and C.P. Slomp. 2020b. Removal of Phosphorus and Nitrogen in Sediments of the Eutrophic Stockholm Archipelago, Baltic Sea. *Biogeosciences* 17: 2745–2766.

Wiggin, B., C. Fornoff, and P. Kim, eds. 2020. *Timescales Thinking Across Ecological Temporalities*. Minneapolis, London: University of Minnesota Press.

CHAPTER 3

Technologically Sensing the Underwater World

The previous chapter mainly engaged the underwater world by bringing sediment core samples up from the water. These samples were then analyzed in experiments to determine what they could tell about the underwater world and its relationships with the terrestrial one. It was shown that results from aquabiopolitical relations not only end up in the sediment, but create situations wherein microbes in the sediment, such as methanogens, actively work in ways that could have considerable effects on humans above water. To further investigate these aquabiopolitical relations this chapter will examine some of the technological devices that allow scientists to study the seafloor and the water between the bottom and the surface. We will look into what technological and infrastructural affordances the research vessel *Electra* offers scientists in their quest to learn about the underwater world.[1] By "affordances" I follow Tim Ingold's (2018)

[1] "Affordance" is a term that was coined by James J. Gibson (1979) that has inspired many approaches and renderings since. However, as noted by Isis Chong and Robert W. Proctor (2020), most usages today differ from Gibson's definition of the term, although without making the difference explicit. The way I use and understand "affordance" in this chapter is similar to that of Donald Norman, in what he labeled "perceived affordances" (Norman 2002). He emphasized that although affordances might be present in the environment/world, not everyone will see or use them in the same way. This way of thinking is similar to that of Keane (2018) who argues that projects and knowledge about the world make certain affordances visible and available.

R. Rodineliussen, *Underwater Worlds*, Sustainable Development Goals Series, https://doi.org/10.1007/978-3-031-63370-6_3

definition as parts of nature that offer humans or other-than-humans opportunities for action. Thus, I view the instruments on *Electra* as affordances in that they offer ways for scientists to engage the underwater world. As Melody Jue (2020: 3) explains with regards to the marine sciences:

> Today, computer technologies perform the work of signal transduction, converting acoustic measurements into visual renderings. Rather than being immediately present to us, the (deep) ocean emerges as an object of knowledge only through chains of mediation and remote sensing—measurements that allow us to build up imaginative pictures of what life in the ocean is like.

However, these instruments only serve as affordances because of the scientists' technical knowledge about the instruments onboard the vessel and their perceived understanding of the world below water. To initiate an understanding of how the instruments on *Electra* turn into affordances for scientists, we will look into the material infrastructure for conducting scientific maritime research, that is, the vessel itself.[2] First we consider the construction of the vessel, her history, the technical equipment she carries, and her role among scientists at Stockholm University. Second, we examine how *Electra*, with all her technological devices, opens a way for scientists to relate to the underwater world. Here I follow Cymene Howe in her definition of some technologies as translators between the environment and humans. Accordingly, the research vessel can be seen as an "empowered other-than-human participant" that facilitates a relational understanding between the terrestrial and marine world (Howe 2019: 74). This is an attempt to rearrange the prevailing terrestrial focus employed when humans think about water (Jue 2020; Raycraft 2020).

To illustrate these technological affordances, we join *Electra* on a training mission with an oceanographer and his students. Rather than conducting research, the task on this trip was to help a marine archeologist from the Vasa Museum in Stockholm to locate one of the Vasa ship's lost cannons. This exercise puts many of the vessel's capacities to the test. We will remain onboard *Electra*, in waters close to the Stockholm city center, mainly moving back and forth across the so-called Vasa pit. The pit is

[2]Within anthropological literature, infrastructure has, especially during the last decade or so, been a key topic of interest (Anand et al. 2018; Björkman 2015; Carse 2012, 2014; Hetherington 2019; Larkin 2013; Schwenkel 2015).

located where the *Vasa* sank in 1628, only a few meters offshore, outside the small island of Beckholmen and next to the larger island of Djurgården. The *Vasa* would have been the largest and most heavily armed warship of her time in the Baltic Sea.[3]

The chapter starts with the material and historical aspects of *Electra* in her capacity as a research vessel. When we have engaged the technologies in their material capacity, we move on to an ethnographic description of when they are used before continuing to what type of relational understanding of the underwater world they offer. We will also discuss the relationship between *Electra* and her crew, how they together constitute the infrastructure needed for research.

The R/V Electra af Askö

R/V Electra af Askö is the newest and most well-equipped research vessel operated by the Baltic Sea Center at Stockholm University. R/V stands for research vessel, and *Electra af Askö*, her name, is a tribute to the late Lars Silén, internationally known for his work on moss animals. The most common moss animal in the Baltic Sea is called *Electra Crustulenta*. Silén also founded Stockholm University's first marine station on Askö in 1961, hence the vessel's name. *Electra* was ready for service in 2016 and is one of the world's most well-equipped research vessels of her size. The ship's design is a collaboration between scientists and crew, making *Electra* especially well-suited for research. The vessel is 24.3 meters' long and 7 meters' wide, it carries a two to three person crew, and as many as 35 passengers for a day trip, fewer for overnight stays. Another of *Electra*'s attributes is her flat bottom, reaching only 2 meters below the surface, which means sampling and other tests can also be done in shallow waters. This is a unique feature that broadens the ship's research capabilities.[4]

Research teams, authorities, and course leaders must apply to obtain space and time on *Electra*. Universities pay the least rental cost, which in 2021 was 1300 SEK an hour for day expeditions and 18,000–26,000 SEK per 24 hours, depending on the number of crew members required on an expedition. This cost generally comes from research grants and may limit the time researchers can spend on *Electra*. For example, the expedition

[3] https://www.vasamuseet.se/en/vasa-history.

[4] Additional reading about *Electra*: https://www.su.se/ostersjocentrum/english/infrastructure/r-v-electra/facts (June 30, 2021).

described in the previous chapter was for five full working days. It had two crew members and cost about 90,000 SEK ($10,500).

Electra's state-of-the-art technologies makes her a one-of-a-kind research vessel. The Baltic Sea Center describes her instrumentation as follows:

> The instruments include a multi-beam echo sounder for detailed mapping ofthe seabed; a penetrating echo sounder that can see into sediments more than100 metres deep; an echo sounder for mapping biology in the water column; aCTD probe that measures conductivity, temperature and depth; and an acoustic Doppler sonar for measuring ocean currents.[5]

These instruments allow *Electra* to engage with water on a detailed level from the surface to below the sediment. By utilizing these instruments, scientists can discover what is hiding below the surface. Furthermore, for example, the multi-beam provides a map of the seafloor and can also find things in the water column between the vessel and the bottom. It can, for example, locate bubbles of methane gas that give indications as to where samples ought to be taken. *Electra* is also equipped with a 31-square-meters laboratory where initial analyses can be conducted, as described in the previous chapter. There is also a control room with a multitude of screens where information from the different instruments appear in real time.

Treasure Hunting

Onboard the *Electra* I met Mathias, the mechanic, whom I had gotten to know during the last expedition, and we greeted warmly. He asked me if I would help him out a bit today, as there would be many students on board and he might need a hand with some of the instruments. I assured him that I would help in any way I could. Then Thomas returned with Martin Jakobsson, a well-known Swedish oceanographer who had come back from an Arctic expedition the day before, looking a bit tired. He was followed by his technician Björn, and the students. Thomas introduced everyone and took the group on a quick tour of the *Electra* before going above to rouse *Electra* from her sleep. At this moment, one more person

[5] https://www.su.se/polopoly_fs/1.362464.1513355146!/menu/standard/file/Electrafolder.pdf.

walked onboard, and it was the marine archeologist from the Vasa Museum, Jim Hansson. He looked utterly enthusiastic and greeted everyone with a big smile.

Martin started instructing his students as soon as we all had come onboard. When the archaeologist joined, Martin quickly interrupted himself, greeted him, and continued his instructions. He explained to the students that they would use a number of the tools offered on *Electra* to experience what they can do and how things work in practice. Then he turned toward the archeologist with a smile, and said, "But rather than just looking at the seafloor, we will help searching for a missing cannon. Jim, please tell us more." Jim began:

> Hello everyone! My name is Jim Hansson, and I am so excited to be here today! We will use the technology on *Electra* to look for the last missing cannon from the *Vasa*. Originally, there were 64 cannons on the ship. Three are in the museum as we speak and were retrieved in 1961 when the ship was brought up. However, during the 1600s, after the *Vasa* sank, 58 cannons were recovered, but the 59th dropped back into the water by accident, according to the archives. It is the 59th cannon that we are looking for today. A real treasure hunt!

Electra quickly reached the place where the *Vasa* sank, and Thomas stopped the ship. Martin started to explain what we would use the different instruments for. The tool for measuring Conductivity, Temperature, and Depth (CTD) would take samples from the water column above the pit to understand the water in a specific location.[6] Martin defined the CTD as "the main oceanographic tool." Sediment samples would also be taken to calculate how much sediment may have ended up on top of the cannon. This information would be important for the museum's divers who would go down to dig for the cannon. Jim explained:

> When we dig in the sediment, all visibility disappears and we dig as if we were blind. This makes it even more important to have as much information as possible before we begin. Even though we do not know the cannon's

[6] Technically, the CTD measures Conductivity, Temperature, and Depth. It does not take water samples. It is the rosette situated around the CTD that take the water samples. However, onboard the vessel, the whole instrument went under the name of CTD. For further reading, see: https://polarisen.se/forskare/forskar-mellan-himmel-och-hav/ctd-rosett/.

exact position, we can at least know how deep to dig before trying another spot.

After explaining how the CTD would be used and the process of taking samples, Martin discussed the role of sonar on *Electra*. The vessel will move back and forth across the pit and "paint" the floor on several screens (I elaborate on this below), using her sonar to search for the missing cannon. If this had been an "ordinary" expedition, sonar would be utilized to map the seafloor in search of topographic shifts resulting from climatic changes, such as melting glaciers and rising water levels. Today its role would be to locate a historical object lost and buried in the underwater world.

CTD: Knowing the Water

The CTD, seen on the images below, can measure the electrical capacity, salinity, temperature, oxygen levels, and depth of a water column. This information holds varying importance depending on what type of research is being conducted. For some research it is crucial information, and for others, it is more background and context for sediment samples and other tests. However, knowing the conditions in the water that is directly in contact with samples helps to understand what is occurring on the ocean floor (this was discussed in the previous chapter with reference to, among other things, oxygen and cable bacteria).

A CTD is a standard tool on research vessels, and it is a multimodal tool that operates with sensors and physically captures water at requested depths. The CTD is a key tool for analyzing the material composition of water. It measures levels of chemicals and the presence of other things in the water.[7] This set of "data" collected by the CTD allows scientists to know what this "water" consists of. Knowing the composition of the water gives information about what can be expected in the sediment. It can also show pollution patterns if there are chemicals present in the water but not in the sediment, or the other way around. Water is a changing substance, and the CTD makes water less abstract and more understandable at a specific location and time, showing that it indeed is an assemblage of different matter (Helmreich 2011; Schmidt 2019; Linton 2010). When Martin called the CTD the most important oceanographic tool, I understand it as

[7] For measuring water and what it contains, cf. Ballestero (2019).

he also acknowledges the importance of understanding this assemblage of relations that is water (Figs. 3.1–3.2).

The images above show the CTD's cylindrical construction. Within the white skeleton are several tubes with openings at the bottom and the top that can be closed at different depths to capture water samples from various water column layers for analysis. The CTD also has a sensor that calculates data and broadcasts it live to monitors in the computer room on *Electra*. For this reason, the scientists preferred to stay in front of the monitors while the CTD made its trip to the bottom and back. However, the CTD requires some manual handling, and here I was able to assist Mathias. He controlled the machine mechanically, while I helped to guide it in and out of the door on the side of the ship, ensuring that it did not hit the sides of the exit/entrance vault. When I first grabbed the CTD, Mathias told me with a smile, "Careful with that—it cost a million or so." Due to Stockholm University's safety regulations, I had to wear a life jacket attached to the ship with a strong rope. As I was about to begin, Thomas came down to look, and when he saw me in the red life jacket about to push the CTD out, he smiled broadly before going back up to the deck. It must have looked rather funny since the jacket clearly was not meant for a person over 191 cm tall.

The CTD is usually run at every station before samples are taken as a way to ensure that the scientists have all available data for that location. It is also a way to ensure that the station looks the way scientists have

Figs. 3.1–3.2 The CTD on Electra. (Photo: Author)

predicted, increasing the possibility that samples will bring the hoped-for results. One could here think along with Andrea Ballestero (2019: 22) and her notion of "infrastructuralizing." Ballestero uses this idea to describe a process whereby researchers and government officials try to render water in aquifers into a commodity that can be counted and controlled. They do so by running multiple types of tests and making calculations based on the results. In the case of *Electra*, the CTD could be seen as a tool scientists employ to infrasructuralize the sea, or at least the specific area where samples will be taken—trying to regulate the "real world" outside of the laboratory as much as possible. This attempt to make the outside world as controlled as possible recalls the issue of objectivity and experiments conducted within the controlled setting of a laboratory. The Baltic Sea is obviously not a controlled laboratory. However, to provide "data" that is considered reliable, certain levels of control of the environment have to be achieved. The CTD provides a measure of assurance that the area being sampled is representational for the type of experiment being conducted on the samples, hence, that a level of reliability will be reached. Still, precisely where the CTD is sent down, and whether or not the results are enough to make the experiments that follow reliable, depends on personal determinations by the researchers; it is a subjective decision.[8]

Before using the CTD, Martin explained its function for the students. He opened and closed the lids and showed the screens where the data would show when the CTD was in the water. Then he sat down in front of the screens with his students, waiting for Mathias and me to get the machine going. Once the CTD had submerged and begun to descend toward the bottom, Martin and his students followed the activities on the monitors. Different colored lines appeared and moved in a zigzag pattern across the screens, the colors representing different measurements, such as salt or oxygen levels in the water at different depths. Martin traced the lines with his finger and guided the group along with them, ensuring that everyone understood their meaning and importance. He noted that levels of oxygen are declining in the Baltic Sea, portending harsh consequences for marine life. The CTD's measurements assist scientists in following how the issue progresses.

[8] This brings the discussion back to Michel Polyani and his book *Science, Faith, and Society* (1946) and subsequent discussions between social and natural science on objectivity and whether or not it is reachable or even desirable. I discussed this more in Chap. 2.

Painting the Seafloor

When the CTD was retrieved and safely stored, *Electra* began moving again. At this point, Martin called to Thomas and asked him to activate the sonars. "Let's paint the floor for the students." Then he returned to his chair in front of the screens and started to tell the group what would come next (Figs. 3.3 and 3.4):

> Now we will paint the seafloor on these screens. *Electra* will move back and forth, and behind her a colored line will appear on one screen. On the other screen we will see a black and white pattern of the bottom that is called backscatter. When enough lines have been drawn, they will combine to form a visual image of the bottom surface. In the first image, different colors indicate different depths. In the second, nuances of black and white show the hardness of the materials. It will appear like an image, and that is why we call it painting the seafloor. Later we can combine both images to make a 3D map of the seafloor.

Electra moved slowly back and forth across the pit. At first, the screen was black with just a small icon of the ship in white, but as the ship moved in the water, so did the icon on the screen. It was trailed by a colored line that, after moving back and forth across the entire screen, ended up as a colored picture of the seafloor. It looked very similar to a painter and the

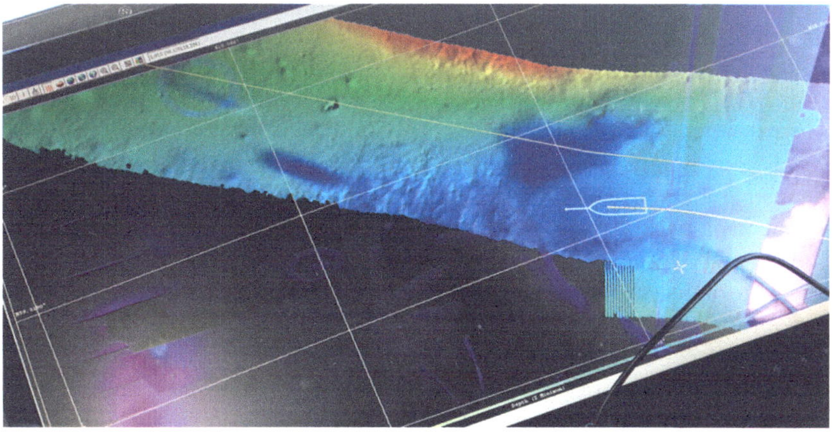

Fig. 3.3 "Painting the seafloor." (Photo: Author)

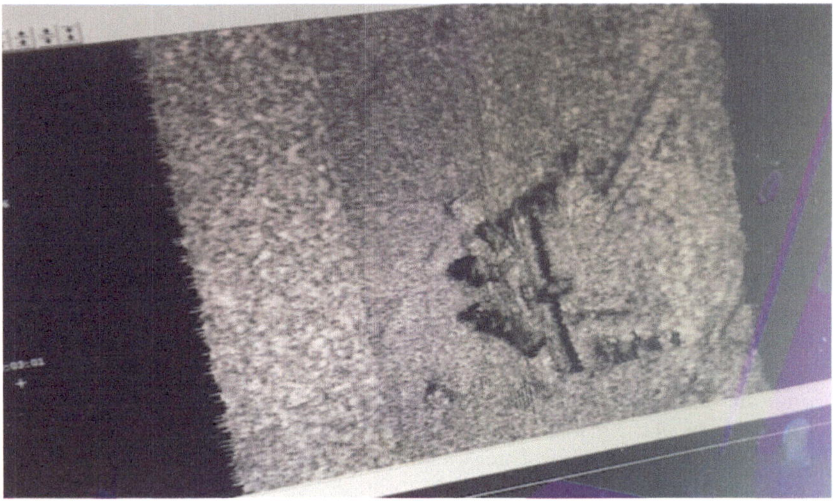

Fig. 3.4 "Painting the seafloor"—backscatter. (Photo: Author)

movements of a brush. *Electra* was painting this image, one line at a time—rhythmically removing the black on the screen and replacing it with a colorful image in changing hues that ranged from warm red to cold blue, with gradations in-between. The different colors indicated variations in depth and types of surfaces. At the same time, the backscatter slowly expanded on the other screen. It was shaped like a long line and showed objects in shades of black and white. Harder objects were indicated by a darker color. As the process unfolded before our eyes, Martin jumped from his chair whenever an object slightly resembling the form of a cannon turned up. Each time he stirred he murmured, "Another piece of wood," or something similar. Everyone's hopes were high that day because the instruments on *Electra* were newer and more precise than those on the marine army vessel that had previously helped the museum in its search. The area where the cannon is thought to lie is not very large, so that even if Martin or Jim did not dare to hope that they would find it today, there was optimism in the air.

This process of looking at the screens as *Electra* painted the seafloor, in a way, reminded me of old-fashioned movie making, where ethnographic films were shot within a single stationary frame. The camera was on a fixed tripod and events played out in front of it (McDougall 2005). Likewise, here on *Electra* the screen is still and events move across it. What is outside the frame is not taken into account. However, what will be shown on the screen has some uncertainty. While the "frame" has been chosen, what may pass within it is not known beforehand. Thus, there is an element of surprise paired with the scientists' power to select where and when to look at the seafloor using the technological affordances on *Electra*. I find this imaginative angle, thinking of the screen as a movie, intriguing because it connects to ideas about framing discussed in the Chap. 1. It points out that there are subjective aspects of marine science: Where Electra looks down on the seafloor is subjectively decided among the scientists.

As we looked at the growing images, Martin commented on what we were seeing. Pointing to the tracks of cruise ships on the seafloor, he said with distress that they are very bad for the marine environment, especially in such shallow waters as this area close to Stockholm. The sound such huge ships make disturbs animals, and the flow of water from the engines hammers the seafloor, destroying what is on top of it. Here three points come forward. Martin's trained eye could interpret the image being painted on the screen (Grasseni 2007). He knew the color codes, what the different lines and shapes indicated, and what to look for in the image. While for the rest of us in the room, it was mostly a colorful and vivid image. Paolo Favero (2020) has argued that images not only represent things, but can have an impact on a viewer's behavior. Thus, the emerging painting of the seafloor has the capacity of affecting the viewer. In this case, for Martin, it created a relational response of anger and an urge to defend the marine world.[9] These tracks of cruise ships along the bottom provide a clear example that the underwater world is a product of both human and other-than-human activities (Hetherington 2019: 2). As Martin told me later when I asked why he uses this painting technique in his research, human-induced climate change alters the environment and this is shown on the seafloor. Temperature changes or rising sea levels create movements along the bottom surface. With his "paintings," he can follow these developments and compare them with historical changes to extrapolate future scenarios. Ocean and seafloors are the largest unexplored areas of the Earth, and as they are crucial for the development of climate change and human life on Earth, it is of vital importance to understand them.

[9] The relationship between image and viewer will be further explored in Chap. 7.

The Relational Affordances of Electra

Humans have long viewed water as a sort of black box where what is thrown inside stays inside and hence out of sight (Reno 2015). Water has been considered something that humans can use as a resource, a place to dump waste or treat in any manner without the effects becoming visible or tangible for us in the terrestrial world. However, the technical equipment on *Electra* makes it possible to look into the underwater world and fathom the proportions of anthropogenic impact. According to Natasha Myers (2019: 125), it is possible to say that the underwater world, like gardens in her writings, "stage moral orders," allowing a moral perspective that looks into what type of world the human-water relationship has created. For example, considering areas where trawlers have vacuum cleaned the bottoms for fish, crushing everything in their way, makes evident what type of life is valued over other lives (Wakefield and Braun 2019). This is exemplified by the way fishing fleets in the Baltic Sea are controlled—not so much for the sake of the fish as for keeping the fish stock large enough to sustain continuous human consumption. The tracks and other evidence of this type of trawling are possible to observe while "painting the seafloor," opening a window to the world below the surface, allowing moral implications, and pointing out victims and villains within an aquabiopolitical process of managing marine life.

By rendering the underwater world visible, *Electra* allows an engagement with it in the form of what Stefan Helmreich (2009: 32) calls a "media ecology, a complex of material, meaningful relations among researchers and their objects of study, relationships structured by techniques of perception and communication." Helmreich further stresses that the message received depends on who the receiver is, and whether he or she is trained to decipher what is rendered visible. This communicative relationship offered by the infrastructural affordances on *Electra* becomes evident if one looks closer at the sonar systems that produce the information that paints the seafloor.

Sonar work by utilizing sound, sending sound waves from the vessel to bounce off the surface of the bottom and, depending on the material encountered, will return waves that register differently and show the type of material below to the scientists. Thus, sand, rock, or wood will reflect soundwaves differently, and therefore it is possible for the scientists on the vessel to get information about what is on the bottom, in addition to depth and other factors. Marina Peterson (2021: 7) describes how "people (and sometimes technologies) listen," showing how electronic devices at and around an airport listen to and assist humans by capturing and translating sound, rather than assessing it as humans do. She also presents sound as a form of touch, that when we listen to something, it is like we touch it from a distance. Similarly, *Electra* utilizes waves of sound to "listen to" the underwater world, working as an other-than-human participant, providing a service to scientists by translating sounds into images (Helmreich 2007; Howe 2019; Besky and Blanchette 2019). Peterson (2021: 5) writes that "what we learn from noise is a rich and dynamic way of attuning toward understanding the atmospheric." I argue that the same is true for understanding the underwater world, because both sound and the underwater world are invisible to the human eye. When sound waves are utilized to paint the seafloor, one invisible sensorium makes another invisibility visible.[10]

I continue to follow Peterson (2021: 23) in her argument that aircraft sounds change the land below by wind and noise. The same argument can be made for the way cruise and trawling ships change the seafloor. The noise they produce affects life below water, as "felt sound," engaging beings in an embodied manner (Campt 2017: 7). Cruise ship engines press water down, and trawlers crush what is in their way, changing the topography of the seafloor. If the same thing happened on land, the outcry to end these practices would have been heard long ago, it is much similar to the destructive human treatment of the Amazonian rainforest. By understanding how aircraft noise affects human life, we can imagine how other-than-humans are affected by the noise of, for example, cruise ships. Even though *Electra* does not listen to underwater noise per se, she does utilize sound to paint a picture of the underwater world where tracks produced by the impact of cruisers and trawlers become visible—the noise accompanied the tracks. *Electra*'s technological affordances use sound to make the underwater world available for humans to engage with, although as visual representations rather than as sound. Thus, when utilizing the instrumental affordances offered by Electra to the scientists, forms of human impacts and "slow violence" (Nixon 2011) on the underwater

[10] Hayward (2010: 581) provides another perspective on the subject of touch-visibility; the "fingeryeye," defined as "the interplay of vision and touch." Hayward builds on the work of Howard C. Hughes (1999) when he "maintains that technologies such as sonar and electroception bespeak a human envy of nonhuman sensoria" (Hayward 2010: 580). The sonar is, then, a fingeryeye providing humans with access to non-human sensoria.

world become visible. The harm done by these vessels to the marine environment is direct, but it is their accumulated harm over longer timeframes that really damage the marine world. The topographic changes that Martin follows with his paintings indicate another form of slow violence, providing indications on effects that already materialized below water but are about to show increasingly also in the terrestrial world as well in the form of sea level rises and more.

Knowledge for the Future

In a long discussion with Björn, Martin's technician, he told me stories about other research vessels, crews, and scientists onboard. He especially talked about the last Arctic expedition that he and Martin just came home from. They had encountered polar bears and wolves, but they also found human traces on land in uninhabited areas. For example, there was one old, abandoned research cabin full of trash that had been left behind. "We had to clean for hours. This will probably end with an official complaint to those responsible for the station. Researchers must clean after themselves!" Our conversation drifted from how personalities are left to bloom in strange ways at sea, resulting in some odd figures, to the importance of sediment sampling. "These samples are usually cut in half. One half is used for experiments now, and the other is stored for later tests. There are some huge sediment storage locations—there is one in Bremen, Germany, for example," Björn explained. I asked, "a bit like the seed valves that exist in the world?" Björn replied, "haha yes, a bit like them. But, after an apocalypse, the seed might be more useful. Don't you think? Haha!"[11]

This conversation made me recall something I was told during my visit to the sediment storage facility in Utrecht: researchers go on more sampling expeditions and bring back more data than they have the time to process. This results in many samples being stored that no one has the time or resources to analyze. It was explained to me that the reason why so many expeditions were carried out is political. When a university or research institute goes to new places, they establish connections with other researchers in that area. It is also a way to show that they care for not just, the Baltic Sea, for example, but for water all over the globe. It is a politics of inclusion played out using research activities and money. This is similar

[11] For a thoughtful and intriguing discussion on seeds, see Karlsson and Rabo (2021), and for "seed valves," see Rabo (2021) more specifically.

to how Brian Larkin (2013: 334) explains "patron-client" relationships, that according to him, guide infrastructural processes. This relationship shows, according to Larkin, the important role political relations play in determining who gets the contract to construct the infrastructure, more so than the quality of the infrastructure itself or the actual delivery of the end-product. "This is why roads disappear, factories are built but never operated, and bridges go to nowhere" (Larkin 2013: 334). This bears a resemblance to why sediment samples are collected and stored indefinitely. The political relations come prior to the actual scientific need for these samples. Political power or goodwill is being acted out. Returning to the sediment storage in Utrecht and finding ourselves surrounded by row after row of sediment cores from all over the world, only to discover that only some samples from each expedition are ever analyzed. It is then when one realizes the magnitude of this knowledge sampling project, where knowledge is stored in the cores on the shelves should they ever be needed. However, as mentioned in Chap. 2, the decision to analyze a sample or not rests with the scientists (Telesca 2020). Hence, which cores to analyze, and from what location, is a decision in the hands of the scientists, the university, or even those funding the research.

Thus, we see that certain individuals can determine what we know about the world below water, and when we learn about it. Industry leaders might not be keen to find out that the water they discharge is harmful, and governments might not want to know exactly how bad the situation is for fish in seas controlled by national fishing fleets. Knowledge is power, but deciding not to know can sometimes be even more powerful, especially if it allows one to keep doing business as usual. This appears to be the driving force behind the aquabiopolitical relations considered in these pages. Authorities and industry only acknowledge that the marine world needs help when the problem is made so visible that they can no longer ignore its presence.

ELECTRA *AND* HER CREW

As I continued speaking with Björn, he talked to me about *Electra's* captain. Björn told me,

> Thomas is one of those captains who not only do their job but who truly love science. He is as curious as the scientists themselves to find good data. This leads to him almost doing the work for the scientists. He can calculate

> where it is better to take samples and when to use the CTD, how to paint the seafloor, and much more. He is a truly incredible captain to have on a research vessel!

This corresponds with a statement by Thomas mentioned earlier that all the scientists joining him on *Electra* seem to enjoy the experience. He explained that *Electra* is a fantastic and well-equipped research vessel. However, from Björn's comment, it is clear that Thomas play a key role in *Electra*'s success. Thomas took part in designing *Electra*, and Mathias, the mechanic, supervised the process of building her at the shipyard. They are both more than workers on a vessel; they are as one with the vessel; they know the ship's every detail and are as one with *Electra*. I spent quite some time talking to Thomas and Mathias about the construction of *Electra*, and they emphasized that she was designed for conducting research, more than for the convenience of her passengers. This objective was agreed upon since expeditions on *Electra* were never to be for long periods of time, so it is less crucial that the ship is "nice to be on" physically than that she has outstanding capacities for research. Both Thomas and Mathias knew exactly how to function on the ship to make the most out of *Electra*.

What comes forth in this description, and sections above, is not a captain and a skipper working on a research vessel, but instead, a captain, a skipper, and a research vessel that work together to engage the underwater world, receiving help from the skillsets of researchers joining them onboard. Together, one could argue, they become what Helmreich describes as a cyborg (2007), human and a technology working as one. The cyborg Helmreich discusses is a submersible called Alvin, whose "work" is to capture and translate deep water sounds. Helmreich builds his idea on Donna Haraway's *A Cyborg Manifesto* (1991: 149), where she defines a cyborg as, "a cybernetic organism, a hybrid of machine and organism, a creature of social reality as well as a creature of fiction." The infrastructure that *Electra* represents provides services to scientists, but she needs her human collaborators (captain and mechanic) to function correctly. If we keep to Helmreich's focus on sound, we can return to the discussion above on how *Electra* utilizes sound waves to touch and paint the seafloor. The vessel thus functions as a dimension between human and seafloor, becoming the link that allow humans to engage with the bottom far below. For this reason, one could argue that Thomas, Mathias, and *Electra* together are the infrastructure, in the form of human and other-than-human collaboration: a cyborg. However, albeit finding the creativeness inherent within the cyborg idea inspiring, it feels a bit far-fetched in the case with *Electra*. In Chap. 6 we will see another instance where technology and human body actively work together to keep the human alive in an environment where it cannot survive on its own, and where the idea of the cyborg may be even more appropriate. While the instruments on *Electra* do not directly interact with the human body to become as one unit, they do offer ways for humans to look into the underwater world. This becomes possible via the knowledge of captain, skipper, and scientists. Following Ingold (2018) and Keane's (2018) definitions, *Electra* offers *affordances* to certain humans capable of noticing and utilizing those affordances to engage the underwater world.

A Window, but No Cannon

On several occasions, what Martin believed to be the lost cannon from the *Vasa* turned out to be a large piece of wood or some other piece of material. When the area around the Vasa pit had been fully mapped by the sonar, no cannon was located. The elusive cannon must have been several meters down, buried below layers of sediment and debris—if it is still there. The sonar on *Electra* can see/listen into the sediment, but in this case it did not penetrate deep enough, or was not precise enough. However, Jim and Martin seemed pleased with the results anyway. "We got good data, and we will analyze the sediment cores together with our measurements from the sonar to try and calculate how deep down the cannon can be," Martin explained. Jim continued by saying that he always thought that the end solution would be for divers to go down and dig by hand, but that the data obtained would be very helpful in directing that work. The search for the cannon will continue.

This chapter has examined a number of affordances on *Electra* that allow scientists to engage the underwater world. We have followed a hunt for the *Vasa*'s missing cannon as a way to demonstrate the technological tools available on *Electra*, and have seen how *Electra* opens up a window to the underwater world to make anthropogenic mistreatment and impact visible. The aquabiopolitical process that is playing out shows that although human actions vis-à-vis the seas are not always visible to the human eye, they are devastating nonetheless.

References

Anand, N., A. Gupta, and H. Appel, eds. 2018. *The Promise of Infrastructure*. Durham, NC: Duke University Press.

Ballestero, A. 2019. The Underground as Infrastructure? Water, Figure/Ground Reversals, and Dissolution in Sardinal. In *Infrastructure, Environment, and Life in the Anthropocene*, ed. K. Hetherington. Durham and London: Duke University Press.

Besky, S., and A. Blanchette, eds. 2019. *How Nature Works. Rethinking Labour on a Troubled Planet*. Albuquerque: University of New Mexico Press.

Björkman, L. 2015. *Pipe Politics, Contested Waters: Embedded Infrastructures of Millennial Mumbai*. Durham, NC: Duke University Press.

Campt, T. 2017. *Listening To Images*. Durham and London: Duke University Press.

Carse, A. 2012. Nature as Infrastructure: Making and Managing the Panama Canal Watershed. *Social Studies of Science* 42 (4): 539–563.

———. 2014. *Beyond the Big Ditch: Politics, Ecology, and Infrastructure at the Panama Canal*. Cambridge, MA: MIT Press.

Chong, I., and R. Proctor. 2020. On the Evolution of a Radical Concept: Affordances According to Gibson and Their Subsequent Use and Development. *Perspectives on Psychological Science* 15 (1): 117–132.

Favero, P. 2020. *Image-Making-India*. London: Routledge.

Gibson, J. 1979. *The Ecological Approach to Visual Perception*. Boston, MA: Houghton Mifflin.

Grasseni, C., ed. 2007. *Skilled Visions*. New York, Oxford: Berghahn.

Haraway, D. 1991. Science, Technology, and Socialist-Feminism in the Late Twentieth Century. In *Simians, Cyborgs, and Women: The Reinvention of Nature*. New York: Routledge.

Hayward, E. 2010. Fingeryeyes: Impressions of Cup Corals. *Cultural Anthropology* 25 (4): 577–599.

Helmreich, S. 2007. An Anthropologist Underwater: Immersive Soundscapes, Submarine Cyborgs, and Transductive Ethnography. *American Ethnologist* 34 (4): 621–641.

———. 2009. *Alien Oceans. Anthropological Voyages in Microbial Seas*. Berkeley: University of California Press.

———. 2011. Nature/Culture/Seawater. *American Anthropologist* 113 (1): 132–144.

Hetherington, K., ed. 2019. *Infrastructure, Environment, and Life in the Anthropocene*. Durham and London: Duke University Press.

Howe, C. 2019. *Ecologics. Wind and Power in the Anthropocene*. Durham: Duke University Press.

Hughes, H.C. 1999. *Sensory Exotica: A World beyond Human Experience*. Cambridge, MA: MIT Press.

Ingold, T. 2018. *Anthropology. Why It Matters.* Cambridge: Polity Press.
Jue, M. 2020. *Wild Blue Media. Thinking Through Seawater.* Durham: Duke University Press.
Karlsson, B., and A. Rabo, eds. 2021. *Seedways. The Circulation, Control and Care of Plants in a Warming World*, Series: KVHAA Konferenser 104. Stockholm: Kungliga Vitterhetsakademin.
Keane, W. 2018. Perspectives on Affordances, or the Anthropologically Real: The 2018 Daryll Forde Lecture. *HAU: Journal of Ethnographic Theory* 8: 1–2.
Larkin, B. 2013. The Politics and Poetics of Infrastructure. *Annual Review of Anthropology* 42: 327–343.
Linton, J. 2010. *What Is Water? The History of a Modern Abstraction.* Vancouver: University of British Columbia Press.
McDougall, D. 2005. *The Corporeal Image.* Princeton: Princeton University Press.
Myers, N. 2019. Five. From Edenic Apocalypse to Gardens against Eden: Plants and People in and after the Anthropocene. In *Infrastructure, Environment, and Life in the Anthropocene*, ed. K. Hetherington. Durham: Duke University Press.
Nixon, R. 2011. *Slow Violence and the Environmentalism of the Poor.* Cambridge: Harvard University Press.
Norman, D. 2002. *The Design of Everyday Things.* New York, NY: Basic Books.
Peterson, M. 2021. *Atmospheric Noise. The Indefinite Urbanism of Los Angeles.* Durham and London: Duke University Press.
Polyani, M. 1946. *Science, Faith and Society.* Chicago and London: The University of Chicago Press.
Rabo, A. 2021. Missing the Seeds in Syria. In *Seedways. The Circulation, Control and Care of Plants in a Warming World*, ed. B.G. Karlsson and A. Rabo. Series: KVHAA Konferenser 104. Stockholm: Kungliga Vitterhetsakademin.
Raycraft, J. 2020. Seeing from Below: Scuba Diving and the Regressive Cyborg. *Anthropology and Humanism* 45 (2): 301–321.
Reno, J. 2015. Waste and Waste Management. *Annual Review of Anthropology* 44: 557–572.
Schmidt, J. 2019. *Water. Abundance, Scarcity, and Security in the Age of Humanity.* New York: New York University Press.
Schwenkel, C. 2015. Spectacular Infrastructure and Its Breakdown in Socialist Vietnam. *American Ethnologist* 42 (3): 520–534.
Telesca, J. 2020. Fishing for the Anthropocene. In *Timescales Thinking Across Ecological Temporalities*, ed. B. Wiggin, C. Fornoff, and P.E. Kim. Minneapolis, London: University of Minnesota Press.
Wakefield, S., and B. Braun. 2019. Oystertecture: Infrastructure, Profination, and the Sacred Figure of the Human. In *Infrastructure, Environment, and Life in the Anthropocene*, ed. K. Hetherington. Durham: Duke University Press.

CHAPTER 4

Slow Violence and the Plastisphere

One of the most visible effects of aquabiopolitical relations, plastics in the aquatic environment, has taken the form of a human-made ecosystem, the so-called *plastisphere*. Marine microbiologist Linda Amaral-Zettler coined the term in 2010.[1] It describes a thriving ecosystem in the marine world in which organisms live and evolve among plastics. This new ecosystem embraces what ordinary ecosystems do, including everything from prey and predators to large and small animals or organisms. Its unique aspect is that the whole ecosystem is made from plastics, that is, anthropogenic waste.[2] "It's like the anthropocene for marine biology," according to plastics pollution scientist Hank Carson (de Wolff 2017: 24).

[1] https://www.theguardian.com/environment/2021/aug/11/welcome-to-the-plastisphere-the-synthetic-ecosystem-evolving-at-sea.

[2] I use the wording "plastics" over "plastic," emphasizing that there is not only one type of "plastic" out there. As noted by Mckay et al. (2020: 2): "'Plastic' is a catch-all category for a group of petroleum-based polymer materials: plastics. For a global public, 'plastic' is found everywhere, both in domestic and industrial settings, but also as a material in the environment." Between 80 and 368 million metric tons of plastics are estimated to be produced globally every year (Geyer et al. 2017; Plastic Europe 2016; Statista 2021). From this amount, approximately 40% is made into single-use products (Plastic Europe 2016). It is estimated that 8300 million metric tons of plastics have been produced up to today, with at least 60% ending up in landfills or as environmental pollution (Geyer et al. 2017).

R. Rodineliussen, *Underwater Worlds*, Sustainable Development Goals Series, https://doi.org/10.1007/978-3-031-63370-6_4

Amaral-Zettler suggests that the plastisphere has existed since at least the end of WWII, when plastics became widespread on a commercial scale, and the volume of the ecosystem continues to expand rapidly. With this expansion, scientists have grown increasingly concerned about the possible toxic impact of plastics on the marine environment. When marine organisms consume or live among plastics, the potential toxicity within plastics can enter the body of the animals. This could, in turn, impact human health when these animals—fish, for example—are consumed as food. Such impacts are now the subject of research. One study from 2021 that focused on degrading car tires (rubber/plastics) showed that these particles effectively kill salmon in water (Tian et al. 2021). Another recent study has found microplastics present in newborn babies (Carrington 2020; Ragusa et al. 2021), demonstrating that plastics travel not only within the environment but now also into human bodies. Hence, separating plastics and bodies, living and dead matter, is becoming increasingly difficult (de Wolff 2017).

In this chapter we will look briefly at the history of plastics and consider the plastisphere's impact on the underwater world. It is of increasing importance to consider the role of plastics, especially since plastics in water not only are now a matter of waste, but have become habitats full of life. The discussion then moves on to governmental and institutional efforts to control plastics and prevent them from reaching the marine environment. We begin with SIWI World Water Week, where UN officials and heads of environmental policy from governments around the world confront the issue of plastics. From this, we move on to a description of how water is cleaned from waste after human use as we tour a wastewater treatment plant in Stockholm. We conclude with a visit to the SINTEF research institute in Trondheim, Norway, where we will meet plastics scientists seeking to understand the toxicity of plastics in relation to life on and with plastics in the marine environment. This group also is working on a possible solution for removing microplastics from water in wastewater treatment plants using an other-than-human: that is, jellyfish bodies.

Throughout the chapter I employ the concepts of aquabiopolitics and "slow violence" (Nixon 2011) to illuminate the way human practices have created a situation in the marine world that has now raised alarm as it poses a danger to human health as well as the health of other living beings and the wider ecosystem. As this impact is becoming increasingly visible, action is being taken on governmental and organizational levels to actively work to reduce plastics in the marine world.

Plastics and the Plastisphere

The history of plastics goes back to the 1500s, starting with the earliest attempt to make plastics from cheese by the German alchemist Bartholomäus Schobinger. The first type of plastics that could be put to practical use was called "parkesin" and was developed in the later 1800s by the English chemist Alexander Parkes.[3] However, although parkesin could be put to use, it was not suitable for mass production. As a result, experiments to make a more production-friendly plastic were conducted during the early 1900s, resulting in the creation of plastics using petroleum and natural gas. With WWII the production of plastics increased exponentially, and this continued after the war. From then on, the uses of plastics exploded, and that escalation has never ceased. Plastics came to occupy more and more space in the homes of people, with numerous products starting to be fabricated with plastics, often as one-time single-use items. At first, plastics were seen "as an environmental good" because the material mimicked capacities of other so-called natural or animal-based products. The idea was that plastics could assume their function while leaving nature untouched (Liboiron 2021: 3). The increasing amounts of plastics in use soon became visible in the environment as they were discarded and accumulated as plastics waste (Pathak 2020: 2).[4] Much of these plastics ended up in the marine environment, where most sank to the bottom. "Plastic moves with wind and flowing water, connecting parts of the terrestrial and freshwater biospheres, moving between them as well as into the oceans" (McKay et al. 2020: 2). The part that stayed afloat started to travel with the water: this is what later was termed the plastisphere. Amaral-Zettler describes the plastisphere as follows:

> Plastic probably has a far larger influence on our life than we think. An increasing amount of plastic waste is floating in the seas and ocean. You can see that when you walk along the high-water mark. If you examine such a piece of plastic from the beach carefully, then you can sometimes see that it is entirely covered by a layer of bacteria, algae and other living organisms. They have mainly seen an opportunity in that plastic; a surface to attach to. Just like the thin layer of life on the outside of the earth, which is called the biosphere, I see the layer of life on a piece of plastic as the plastisphere.[5]

[3] https://www.tekniskamuseet.se/lar-dig-mer/industrihistoria/plasthistoria/.

[4] https://atervinningsblogg.se/vad-kan-du-om-plast/.

[5] https://www.nioz.nl/en/about/organisation/staff/linda-amaral-zettler.

The plastisphere is thus a moving ecosystem, constantly growing, that circulates in the water on the planet. It is concentrated in the plastic gyres in the Atlantic, but is present in all water where plastics float. Organisms attach to and live on these pieces of plastics.[6] Some can even consume it: In 2016 a group of Japanese scientists discovered bacteria that can eat plastics. These bacteria were called *Ideonella sakaiensis*, and they break down plastic using an enzyme that allow it to separate the plastic polymers into monomers that the bacteria can absorb (eat). At first little attention were given to *Ideonella sakaiensis*, but with escalating plastics pollution in the world, more and more interest has been placed on the possibility that these bacteria can solve the plastics issue by consuming the plastics we use.[7] However, scientists are skeptical that it will be possible to grown enough bacteria, or make them efficient enough, to consume the enormous amounts of plastics produced yearly, not to mention all that which is already left in the environment.[8] Hence, it seems like the solution to plastics in the marine environment has to be a change in how humans act with plastics and plastics waste—we need a human solution for a human made problem.

[6] The plastic gyres are specific locations where plastics have accumulated in a rotating body of water. Here plastics keep on accumulating while degrading into smaller pieces. The largest gyre is now calculated to be roughly the size of Texas.

[7] Bradshaw (2022) has investigated *Ideonella sakaiensis,* looking into "how microorganisms might express their own directives, preferences, and constraints on the research process, and how, in turn, we might listen and be directed by them." In this study, Bradshaw highlights how microorganisms, like *Ideonella sakaiensis*, can be seen as collaborators and co-producers of knowledge about ways to combat the escalating issue with plastics in the environment. In this attempt, Bradshaw points out the importance of bridging the divide between the hard and soft sciences in order to find new ways of engaging with other-than-humans to create other possibilities for the future.

[8] For further reading about *Ideonella sakaiensis* see: https://www.sci.news/biology/ideonella-sakaiensis-bacterium-can-break-down-metabolize-plastic-03693.html and https://www.theguardian.com/environment/2023/sep/28/plastic-eating-bacteria-enzyme-recycling-waste.

SIWI World Water Week: And the Source-To-Sea Approach

SIWI stands for the Stockholm International Water Institute. It is largely financed by Swedish international aid money, but also inter-governmental actors who are not from the private sector. The Institute explains its presence and role as follows:

> We trace our roots back to 1991 and the launch of the Stockholm Water Prize and the Stockholm Water Symposium. Three decades later, the Stockholm Water Prize is the world's most prestigious water award and the symposium has evolved into World Water Week, the leading international water conference.[9]

When I attended the World Water Week in the summer of 2019, the theme was plastics pollution, and the invited speakers and participants all sought joint solutions to the escalating issue. Peter Thomson, the UN Secretary-General's Special Envoy for the Ocean, was one of the key presenters. I sat in the audience during one session when he was scheduled to speak. The auditorium was full of chatter and low-voiced conversations, but they all went silent when Thomson took the stage. He walked out to the microphones in the middle with comfort in his steps and a broad smile. That year's SIWI World Water Week was the first time the theme concerned both fresh- and saltwater, including the ocean, and Thomson began his talk with a note on the relationship between fresh- and saltwater. He impersonated the saltwater and posed himself with the "salt-people" in contrast to those working on the "fresh" side of the spectrum:

> What you, freshwater, give us, the ocean, is pollution and waste. In return we provide you with opportunities for shipment and food—lots of food, and we even provide you with recreational opportunities. Well, lately, we have started to send your waste back to you, inside the fish you eat. We need to talk!

Thomson thus dramatized the fact that waste, including plastics, reaches the oceans via rivers and waterways from inland societies. This stressed one of the week's key themes, namely, to further launch and stress the importance of the source-to-sea (S2S) approach, which addresses the

[9] https://siwi.org/who-we-are/history/. The citation was collected from this website in 2020.

issue at its source and seek to prevent waste entering the environment from the beginning.[10] Calculations suggest that about 8 million tons of plastics enter the marine environment from terrestrial sources yearly. For this reason, government officials look to reverse this trend and stop the waste at its source, the point where it first leaks out of the "system" and into the environment.[11] This can entail reducing the number of one-time products or making waste management practices more efficient. Solutions are context-based, but the overall goal is the same: lessening the impact of plastics pollution. There is currently consensus across the globe that it is more effective to address plastics pollution at the source rather than at the end-destination.[12] This is the core idea behind the S2S approach. However, as will be discussed in Part II, is it not also a convenient and resource effective position for government and industry, to not have to clean the waste already in nature, and instead focus on future leaks? Is it not possible, one could even argue responsible, to do both?

I asked at the SIWI information desk if the S2S approach was meant to be applied across the globe, or if specific areas were to be targeted. I said that I was specifically curious about the Swedish situation. The answer was that SIWI does not focus on Sweden, as the issue with waste is not as acute here. Instead, their concentration is on rivers in Asia, for example, that transport considerable waste into the oceans. However, as I was told, it is not that they believe Sweden to be problem-free regarding waste, but that their funding and orientation is for developing countries.

Geographical location is an important factor in how governments and the public relate to production, consumption, and waste. Emilio F. Moran (2016) describes how the geographical distance between consumers and a product's ultimate environmental impact facilitates certain types of practices. However, he suggests that this state of affairs can be explained by a lack of information. Although this may be true in some situations, Sweden's case seems more akin to what Vincent Ialenti (2020) discusses as a growing distrust in expert knowledge among the educated public. The general public in Sweden knows that waste has no place in nature, and still

[10] https://www.siwi.org/what-we-do/source-to-sea/.

[11] Joshua Reno (2016) provides an intriguing account of how waste "leaks" from within a closed system—a landfill—and into the surrounding environment. The leak can be infrastructural and material failure, but also other-than-human involvement, such as birds picking among the waste and moving it out of the landfill's confines.

[12] A UN report summarizes this standpoint: https://www.unep.org/resources/turning-off-tap-end-plastic-pollution-create-circular-economy.

much waste ends up there. Information is abundant, and in some cases overwhelming, resulting in a sort of "compassion fatigue" (Moeller 1999) that can lead to inaction. When government funds point to the issue elsewhere, it is easier to accept paying to fix a problem "over there" instead of having to actively change daily practices here.

How Plastics Leak Through the Cleaning of Water After Human Usage in Stockholm

Peter Thomson's talk made clear that waste, including plastics, travels from human societies toward the ocean via freshwater rivers and waterways. The S2S approach is meant to address this issue at the source by removing unnecessary objects that may potentially turn into waste from the production line before they have the chance of joining the stream flowing into the ocean. However, many items and substances still move from production to consumers and are ultimately discarded as waste, so that much of it still finds its way into the environment. One of the main collection points for this waste, after human usage, but prior to the water being released back into oceans and lakes, are wastewater treatment plants.

To understand what wastewater treatment plants do to clean water post-human usage, I visited Bromma Reningsverk in 2019, the oldest wastewater treatment plant in Sweden. It is one of the main wastewater treatment plants in Stockholm, but in 2013 it was decided that the plant should be discontinued and efforts are made to redirect the wastewater to the newly modernized Henriksdals wastewater plant. Bromma Reningsverk was constructed in 1934 and expanded in 1966 to include another facility. Today the plant cleans the water used by approximately 290,000 people. It takes 24 hours for the water that reaches the plant to be cleaned and released back into the marine environment. When I arrived at Bromma Reningsverk for the first time I was met by an older employee with kind eyes and a broad smile, whom I will call Bengt. He shook my hand and warmly welcomed me to the plant. Bengt has been there for many years, and has worked at most of the jobs within the plant. We had been emailing back and forth prior to my visit, so he already had a general idea about my project and what I might want to look at. We walked into the plant while making small-talk about my research and he found it interesting that someone with a social science perspective was interested in actually visiting the plant. As soon as we entered, I had an immediate sensorial experience of the plant: there was a mild but distinct odor of sewage in the air. It was rather subtle, so I assume that those working there no longer smelled it, but I felt like it surrounded me in a way that made me think I too would smell like a sewer when I left.

Bengt told me that we would walk through the plant following the route that the wastewater travels as it moves through the plant. The first stop on our walk was the mechanical filter, shown on the left image below. It is a metallic net designed to trap objects larger than 3 millimeters. The net rotates slowly in order to not become overfull. It is effective at removing most large objects from the water. The amount of waste collected by this filter reached 620 tons in 2018. This waste is then burned and used for heating. However, as Bengt explained, many small things pass through the filter and continues onward the marine environment. This is because the filter often breaks, or a tear can open up and let items leak through. Since the filter often breaks down, many repairs are needed, which is something that causes delays in the plant's operation that reduce its capacity to clean wastewater. Due to the rather large amount of waste that passes through the mechanical filter, Bromma Reningsverk has installed a "strainpress," a sort of centrifuge that adds another level of filtration for larger objects in the water. This is important here because that which passes the filter that is not water is now filtered away and turned into "primary sludge" that together with the "biological sludge" (explained below) will be part of that which the plant later sends out to farms as nutrients for agriculture. This allows a circular system where waste is used and reused within human societies, rather than directly emptied out into the marine environment. The sludge is visible on the right image below as it is being loaded on a truck for shipment to a farm.

After the mechanical cleaning the water passes through basins where it is slowed down so that heavy particles such as sand will settle to the bottom and be filtered away. This sand can later also be used as fertilizer on farms as it contains some biological material, or as filler for road construction and similar projects. The water will also be filtered through sand filters that will catch any remaining large particles (Fig. 4.1).

The next step on the water's journey through the plant is the chemical cleaning station. Bengt explained that one of the main things done at this station is to use chemical coagulants, aluminum or iron-based, to remove, among other things, pathogens and organic compounds. They do this by destabilizing the wastewater particles, which then form into larger bodies that can be filtered out of the water. This takes place in big cylinders that

Fig. 4.1 Left: The mechanical wastewater filter; Right: The sludge being loaded on a truck for transport to a farm. (Photo: Author)

spin the water around. The larger clusters will sink to the bottom and are removed prior to the water being allowed to continue onwards (Fig. 4.2).

The final stage of the water treatment process in the plant is the biological cleaning. Bengt explained that the cleaning done in the biological part is largely similar to a process the ocean does itself. It is called nitrification or denitrification. This means that excess nutrients are removed from the water, nutrients that would otherwise add to the eutrophication of the water, which affects, among other things, the oxygen levels in the water and thus leads to dead bottoms. While the ocean can do this itself, it cannot handle the enormous amounts that come from human societies, which is why wastewater treatment plants are a necessity.

To increase the effectiveness of the bacteria that carry out the biological cleaning, oxygen and methanol are added to the water. The bacteria do not die, but multiply in the plant. "We feed them oxygen and methanol to keep them happy so they will work well," Bengt said. Furthermore, he explained that the methanol used comes from fossil sources, which is something the plant is currently working to change. After the biological

Fig. 4.2 Left: Iron sulfates interact with the wastewater; Right: Adding oxygen to the water to increase the activity of bacteria during the biological cleansing. (Photo: Author)

cleaning, the remaining sludge is left to decay in large silos, a process that produces biogas. There is one large blue silo and six smaller in metallic color. In the large blue one the primary sludge will be made to rot while in the remaining six already rotten primary sludge will be mixed with the biological sludge to continue the process together. The sludge is later centrifuged to separate out all remaining water and is then transported to farms. Bengt explained that in the sludge sent toward the farms there can theoretically still be microplastics, because Bromma Reningsverk does not have the capacity to capture microplastics. "This is one of the good things with the new wastewater treatment plant being built in Henriksdal, it will be able to also catch these microplastics" (we return to this). The water centrifuged away from the sludge can contain high levels of nitrogen, sometimes above 10%. Formerly, this water was channeled into the main water current that exits the treatment plant. Now it is first run through a new system called Anita Mox.[13] This system works by directing the water

[13] https://www.veoliawatertechnologies.com/en/solutions/technologies/anita-mox

through a container with a spiral-shaped metallic stick in the middle, and on the metal there is a bacterial culture that interacts with the nitrogen, binding it and removing it from the water. The bacteria used is the annamox bacteria described in Chap. 2.

Plastics Slow Violence: A Theoretical Detour

Plastics leak from societies into the marine environment in numerous ways, and have done so increasingly since the 1950s, despite efforts to contain them. And look to keep on doing so for the foreseeable future. Therefore, I wish to think about this possible violence as taking place in space and time, on a local and global scale and as visible and invisible causes and effects. Barbara Adam pointed toward the importance in thinking of space through time in relationship to in/visibility in her book *Timescapes of Modernity* (Adam 1998: 9), where she writes:

> While space is associated with visible matter and sense data, time is the invisible 'other,' that which works outside and beyond the reach of our senses. This makes time such a pertinent focus for environmental issues .... We are dealing with phenomena where the impacts of actions work invisibly below the surface until they materialise as symptoms—sometime, somewhere ... these industrially produced phenomena and processes are characterised by invisibility and periods of latency.

We can extend this with Rob Nixon's (2011) notion of "slow violence." Nixon defines this type of violence as not associated with a big explosion, a large oil spill, or anything else directly visible to the human eye. Slow violence, according to Nixon, is a haunting type of violence, one that accumulates in the periphery and slowly, silently, creeps upon you and inflicts harm. It is not easy to point out when the harm was created or where it will strike the hardest, but strike it will.

Today we are experiencing a time when plastics in the environment are, in a way, striking back. This is evident if thinking about the example I provided in the introductory chapter: how no less than 43,800 bathtubs worth of waste float onto the Swedish west coast yearly—much of it plastics. These plastics have accumulated below the surface, out of human sight, for decades. It is only now that this accumulated waste starts to show itself, and the harm it can do to humans and other-than-humans become visible.

For example, animals full of, or entangled in, plastics are turning up dead on beaches and elsewhere. Thinking in terms of "slow violence" offers a lens to understand the incremental contamination of the environment by plastics from the early 1900s until the situation we have today. The oceans have slowly been filling up with plastics, and the effects are becoming visible in the enormous amounts that wash up on land. Governments, agencies, and organizations have been forced to begin cleaning up the mess that has been in process for decades. This reaction also brings attention to the fact that humans act when damaging results become visible on land, that is, when we, the humans, can see them. Gauri Pathak (2020: 11–14) has discussed the relationship between visibility, action, waste, and plastics in terms of aesthetic pollution. She argues that much anthropological research has overlooked the aesthetic dimension of pollution, which in the case of plastics in the environment becomes misleading due to the enormous role played by visibility.

I use the concept of aquabiopolitics to show that it is the visibility of plastics (or other waste/pollutants) in nature that now forces governments and organizations into action. It is also by making plastics in nature visible that environmental activists put further pressure on governments to act. Thus, while the slow violence caused by plastics pollution has accumulated over time, it is only when plastics become (or are made) visible that the issue is acknowledged and acted upon. As Adam (1998: 37) argues, the "preoccupation with visibility, materiality, and space is deeply embedded in the knowledge traditions of western societies." Politicians tend not to act before they feel they have a scientific basis to act from, and that can often take a long time to develop.[14] This is why some activists engage in making current events into a "moral punctuation" that circumvents scientific processes of knowledge production and concentrates long-term processes into a current happening that, in a way, forces politicians to act (Ahmann 2018). I return to this in Chap. 7.

[14] This is something discussed by Ialenti (2020: 27, 32–33) as what he calls "the deflation of expertise," by which he means that the public is growing less trusting of scientific knowledge. It often is too slow, and therefore politicians that await scientific results tend to act late. This leads to further changes in trust in public institutions among the population. I return to this in the second part of the book.

Plastics and Jellyfish

To further investigate the relationship between plastics and the marine environment, and to understand more about a newly proposed solution for removing microplastics from wastewater in treatment plants, I traveled to the SINTEF research Institute in Trondheim, Norway, which was founded in 1950. The center runs only on research funding and grants, which creates a climate where all employed researchers must attract grants to keep their positions. This is, in a way, a stressful environment, but as Rasa Slizyte, one of the researchers I visited, explained, it has upsides too. For instance, they can put all their time into research. Compared with the University in Trondheim, where they also need to find time for teaching, here at SINTEF, all time is devoted to research.

I had traveled to visit plastics scientists Andy Booth and Lisbet Sørensen, biotechnology scientist Rasa Slizyte, and political scientist Rachel Tiller. They are all part of a project I call "Jelly-Fix" that utilizes mucus extracted from certain jellyfish to make a microfiber membrane that can be used as a filter in wastewater treatment plants.[15] This filter would effectively catch microplastics that otherwise would pass through wastewater plants and out into seas and oceans. Andy and Lisbet worked on the plastics aspect of the project to understand the role of plastics in the marine environment, what toxicants are released, how and what bacteria live on plastics, and how far plastics can travel in water. Rasa worked on the collection of mucus and chemical characterizing of this material. Meanwhile, Rachel looked into legal and political issues regarding plastics pollution and the possibility of harvesting jellyfish on a larger, more commercial scale, which would be needed to implement this proposed solution for capturing microplastics. In what follows, we will examine the research on plastic waste and the attempt to use jellyfish as a solution for a human-made problem.

Plastics

I joined Lisbet in the laboratory (much of the technology within was similar to the biogeochemistry laboratory in Utrecht described in Chap. 2).

[15] The real name of the project is anonymized as it is still ongoing and the scientists asked me to not name it in writing, although I was permitted to describe their work and their objective.

She walked me toward the more specific instruments they used to analyze the role of plastics. Our first stop was a room containing five rotating carousels with bottles attached. Lisbet saw my puzzled expression and explained (Fig. 4.3):

> These bottles contain water, small plastic bits, and natural microbiota (bacteria). They spin like this to mimic the movement of the ocean. We look into how and if different types of bacteria can attach to and grow on these pieces of plastics. This is important for an understanding of how microbiota can interact with plastic to change its fate and effect in the environment.

In an adjacent room were bottles containing different types of water, fresh and from a fjord. Inside the bottles were pieces of plastics and bacteria. Scientists looked at how the bacteria interacted with the plastics, depending on the type of water. Lisbet further explained that on top of each piece of plastics there was a thin layer, and on this layer, bacteria and

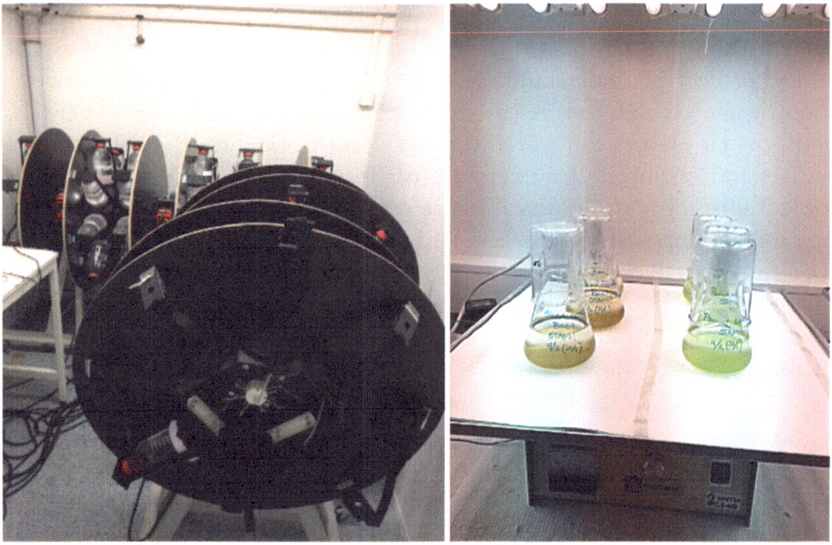

Fig. 4.3 Left: The rotating carousel; Right: Bottles from a study with marine (brown) and freshwater (green) microalgae. This other study looked at the effects of plastic particles and chemicals from plastic particles on these algae, specifically at how it can restrict their growth. (Photo: Author)

other organisms can attach themselves and possibly thrive. In a way, each one of these bottles contained a tiny part of the plastisphere. It is, however, important to note that Lisbet did not mention the idea of the plastisphere herself, but that is how I understand these plastic particles and the layer on top of them in the larger context discussed in this chapter. This experiment allows us to ask if the plastisphere is the same in all water, and it provides possibilities of exploring whether this human-made ecosystem varies across the globe, depending on the local composition of water and types of plastics. Amaral-Zettler's research has shown, for example, that varying colors on plastics indicate amounts of life on that particular piece.[16] This is also something that Lisbet and Andy investigate when they use not only different types of water but also different types of plastics in relation to what type of bacteria they include in their experiments.

Andy brought me back to the topic of slow violence when he pointed out the potential toxicity of these pieces of plastics, while emphasizing that this is still ongoing research, and for this reason, they cannot say for sure how toxic microplastics actually are. However, studies indicate that many types of plastics leak toxicants and can damage both humans and other organisms, especially in an aquatic environment. For example, in one study that Andy and Lisbeth were part of they found:

> Plastic associated chemicals represent a hazard to aquatic organisms through leaching, but potentially also to organisms that ingest or otherwise come into direct contact with plastic. Furthermore, they represent a technological challenge to be overcome for plastic waste handling and recycling, where methodological pipelines to accurately address the presence of such chemicals in plastic waste streams are needed. Toxicity assessment of most plastic materials and microplastics (MP) is typically conducted using simple reference materials that do not accurately reflect the complex plastic materials and particles derived from plastic products that are normally found in the environment. Importantly, a recent review highlights the alarmingly low number of studies that factor in associated chemicals when addressing plastic environmental hazards. (Sørensen et al. 2023)

In the quote above, we find references to the complicated work of removing plastics from the waste stream from societies to the marine environment (as described in the section on Bromma Reningsverk), and also

[16] For a comprehensive review of research regarding the "plastisphere," see Amaral-Zettler et al. (2020).

to the issue of how chemicals react with other chemicals—a topic we return to in Chap. 6. To exemplify, Andy took the example of car tire particles that wash off roads and into water (car tire toxicity is a topic we return to in Chap. 7). He explained that these car tire particles are very toxic, and as mentioned above, studies have shown how they can kill, for example, salmon in water (Tian et al. 2021). Andy pointed out that if looking at the issue from a strict anthropogenic perspective, plastics pose a somewhat diffuse danger, as we do not yet fully know how harmful plastics in human bodies can be. However, in a larger perspective that includes other organisms and the environment at large, plastics are a huge problem, according to Andy.[17] Especially because it is so difficult to collect plastics already in the environment, and increasingly so, as they become smaller and smaller with time. Andy ended his explanation on the note that in deeper waters plastics degrade slower than on land. This means that plastics will stay in the water even longer than elsewhere, continuously leaking toxicants. Thus, viewing plastics in the marine environment as a form of slow violence is not a far-fetched line of thought.

The project's plastics research is also conducted in a cold room holding 5 degrees Celsius. There a buoyancy experiment with plastics in the water was taking place. Inside the cold room, I met an intern from Spain, spending a few months at SINTEF to gain experience, and she was just about to begin the experiment (Fig. 4.4).

When Lisbet had left the room, the intern and I greeted, and she began explaining the experiment for me while working. She explained that the long rectangular tube, shown in the image above, contained ordinary seawater. When the experiment begun a bottle containing plastics particles of varying size, type, and density was added to the water. A camera then took over 1000 frames per second to calculate the speed of the plastics particles falling to the bottom. The enlarged images came up on a computer screen placed on a table at the end of the room. So, while we could observe the particles live, we could also view them enlarged on the screen. We stood in the cold room for close to two hours, observing these particles. Some sank right away, others more slowly, and some not at all. The way these particles act in water is used to calculate how far and how long different types of plastics can travel in water. This knowledge can then be paired with data from the carousel described earlier, providing an understanding of which

[17] For further reading based on work of Andy and Lisbeth on how plastics can affect marine life see: Sun et al. (2024), Hägg et al. (2023), Sun et al. (2023).

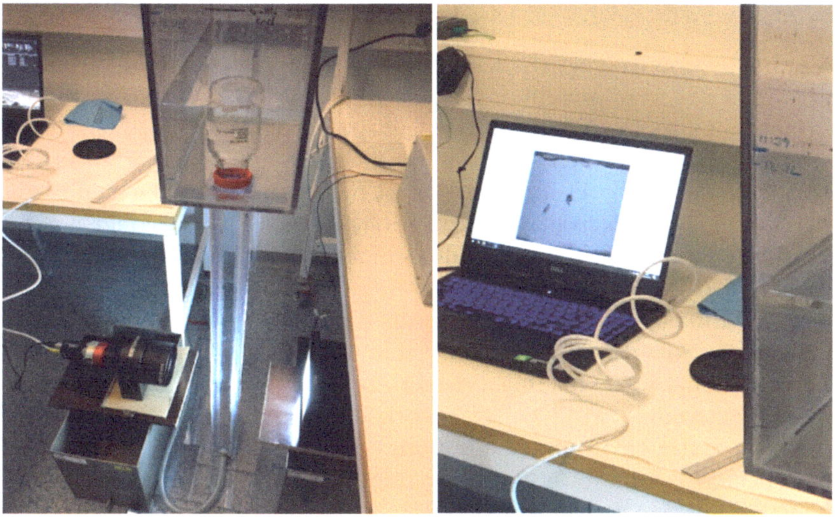

Fig. 4.4 Left: Water tube with plastics particles and camera; Right: Screen showing the upsized images. (Photo: Author)

types and how far bacteria and other organisms can travel on the buoyancy of plastics. This cold room simulated the temperature in lower levels of water. Thus, like the carousels, the room was designed to mimic a real-world scenario as closely as possible. For the two of us, the intern and I, it was freezing cold staying with the experiment observing for two hours. We could not move around much due to the small space and the risk of touching the water tube which in worst case could mean tampering with the experiment. Thus, we stood still, watching, and freezing.[18]

[18] For further reading about earlier studies and the development of the instruments described above, see an overview here: https://www.springerprofessional.de/en/microplastic-fate-and-impacts-in-the-environment/20166256; and for car tire toxicity: https://www.frontiersin.org/articles/10.3389/fenvs.2020.00125/full; and algae toxicity (includes also car tires): https://www.sciencedirect.com/science/article/pii/S0043135419310449?via%3Dihub For the "Silhouette camera": E.J. Davies, P.J. Brandvik, F. Leirvik, R. Nepstad, The use of wide-band transmittance imaging to size and classify suspended particulate matter in seawater, Mar. Pollut. Bull., 115 (2017), pp. 105–114. And for the "Sinking column": O.G. Brakstad, D. Altin, E.J. Davies, M. Aas, T. Nordtug, Interaction between microalgae, marine snow and anionic polyacrylamide APAM at marine conditions, Sci. Tot. Environ., 705 (2020), Article 135950.

Jellyfish

Rasa invited me to meet the jellyfish in the biotech laboratory, although no living ones were present as a new shipment had not yet arrived. Rasa brought me first to a cold room, where she brought out small bags containing a powder of dried jellyfish and mucus. She told me that the primary goal of the Jelly-Fix project is to use mucus to construct a membrane that would catch microplastics. However, as the jellyfish would not be returned to the water, the group also sought to use the remains of the jellyfish so that they would not go to waste. One of the intended uses is as fertilizers for agriculture. These bags she showed me were bound for Germany as a trial run. Rasa explained that in Germany, industry uses jellyfish as raw material for extracting collagen, but they buy them from China. Should these more "local" jellyfish work, it would mean less shipping and thus be a better environmental option. Other ways the jellyfish could be used include as food for human consumption and as an ingredient in cosmetics (Fig. 4.5).

The jellyfish are harvested and brought to the laboratory following a protocol created by the Jelly-Fix project. (1) The jellyfish are caught in a trawl or a dip net. (2) The jellyfish are transported to the laboratory in seawater tanks, within three to five hours. (3) Mucus is collected in the

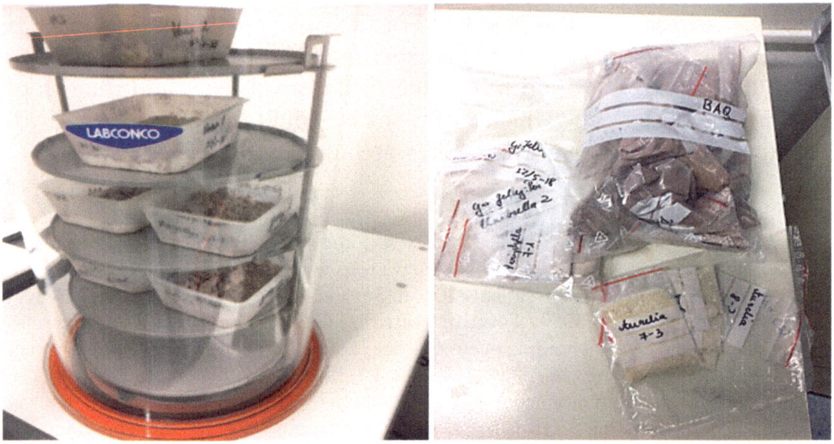

Fig. 4.5 Left: Jellyfish being freeze-dried; Right: Bags containing jellyfish powder. (Photo: Author)

lab. This means that they are "put under stress or stimulation," because it is then that they produce the mucus that is the "part" of the jellyfish needed to construct the membrane filter (I return to the filter below). The stimulation is done by putting the jellyfish in a bottle and gently squeezing and touching them to trigger their mucus production. (4) Then the mucus is filtered through a mesh net, (5) before immediately being stored at –20 degrees Celsius.

After the mucus is collected, the jellyfish are frozen, dried, and ground into a powder, coarse or fine, depending on the intended use. For agriculture, normal grinding is enough, while for cosmetic purposes the powder must be dissolvable in water. In order to make the powder dissolve, an enzyme is added to it that breaks down the structure of the jellyfish, similar to the way our stomachs process food. Ideally, the jellyfish should be freeze-dried, as in the image above (I described freeze-driers in Chap. 2), but Rasa explained that freeze-drying is expensive and is mostly employed for value-added products. For fertilizer, a less expensive technology, drying jellyfish in an oven, is used. The latter is much quicker and cheaper since jellyfish are 85–98% water, which makes the freeze-drying process lengthy. At the time of my visit, Rasa compared the freeze-dried samples with the oven-dried ones to see if the treatment in the oven lost any enzymes or minerals. If not, then they will do all the jellyfish in the oven as it saves much time and energy.

One key question remained at the time of my visit that no one could answer for me. It was about the abundance of jellyfish in the water. Are there enough jellyfish to use them in this way? They must rely on biologists to calculate the amount of jellyfish in the water. However, the general impression is, according to Rasa, that there is an abundance of jellyfish, too many in fact, as they interfere with human-built infrastructures and disturb fishers.[19] But it is not known for sure when and where they are found, and it is hard to obtain permission to fish for them on a large scale. Rachel explained:

[19] Former "jelly-scientist" Juli Berwald's book *Spineless: The Science of Jellyfish and the Art of Growing a Backbone* (2017) gives a comprehensive account of jellyfish as a species. She also provides many examples of human-jelly encounters, as when jellyfish clog water intakes to power plants or when they disturb fishers by hampering the catch of fish. Hence, see this book for further reading on human-jelly interactions and some more general audience type of jelly-knowledge.

> There are many examples of humans believing there are more fish in the water than they possibly can take up. Only to later realize that this is not the case. So we must be very sure before we can start harvesting jellyfish on a commercial scale.

The project is also looking into farmed jellyfish but is unsure if aquaculture can produce the same quality of jellyfish as in the wild. Rasa fears that "domesticated" jellyfish will not be as "good" as wild ones. But she stresses the urgency of answering the jelly-abundance question,

> We are almost ready at stage 2 of the project: how to use the jellyfish. However, without being sure about stage 1: are there enough jellyfish, and can we harvest them? We cannot proceed much further.

Our discussion turned to the topic of how to catch the jellyfish. Rasa told me that from the beginning, they wanted a massive type of jellyfish from deep waters. However, it turned out that some species of jellyfish were very sensitive to light. The group does not have the proper equipment or technique to catch these jellyfish effectively. Thus, they decided to use the much smaller but less light-sensitive *Aurelia aurita*, or as it is commonly known, moon jellyfish. These jellyfish have a diameter of about 5–40 centimeters, and a lifespan of 6–12 months. The moon jellyfish is recognizable by four circles on the top that are visible through its translucent body and these circles often have a purple color. Moon jellyfish are also easier to catch. However, due to their smaller size, greater numbers are needed.[20]

[20] Turning the environment and/or other-than-humans into a resource is part of the modern human becoming. There are a number of studies that illustrate this process of making "nature" into a resource. For example, Tanya Richardson and Gisa Weszkalnys (2014) writing in the introduction to an issue of *Anthropological Quarterly* focusing on the relationship between humans and other-than-humans state that "Natural resource exploitation—as a sustained project of abstracting substances identified as useful, valuable, and natural in origin from their environment—has long played a central role in that continuing human effort to become 'modern.'" Furthermore, Jeremy Schmidt (2019) has shown how, in the US, water has been made into an abstraction so that it can become part of a neoliberal process of using water for economic and political aims in a process he calls "modern water."

Why a Jellyfish Filter?

As described above, wastewater treatment plants use several stations in order to clean different types of waste from water before passing the water on to seas and oceans.[21] In the recently modernized wastewater treatment plant at Henriksdal in Stockholm, a new membrane filter has been installed to catch objects under 3 millimeters in size. It is the world's largest facility of its type, according to the plant's statement. This filter is intended to capture all those particles that have passed through the initial filtering. The plant claims its membrane filter can also remove some bacteria and even viruses from the water.[22] When I discussed this new filter with Bengt from Bromma Reningsverk, the other wastewater facility in Stockholm, I was told that "the membrane filter is very effective. But it also requires a lot of energy. They will need huge amounts of energy to power the filter." Thus, while the filter allows the capture of, for example, microplastics, it provides another stress factor on the environment—increased energy consumption. Bengt further stressed that although the new plant fits the political agenda of being in the forefront of environmental thinking, it might cost more than it saves. One of the main arguments behind building the plant was that it would save money because otherwise the city would have had to expand the old one in Bromma due to the increasing population in Stockholm, which would have been too expensive according to the politicians. However, constructing the new plant has already cost about seven times what it would have to expand the old one, according to Bengt. "It feels like it is mainly a show-off project, hey look at us we are the most modern in the world, type of construction." The filter being developed by the Jelly-Fix project is designed to catch similar particles as the one in Henriksdal, but the Jelly-Fix attempt will be less energy-consuming and will instead use the bodies of jellyfish.

In a study Andy was part of, they write that, "Owing to the increasing evidence that NPs [nanoplastics] present an exposure hazard, there is growing interest in the development of technologies to remove them from

[21] For further reading (in Swedish) on the number of particles coming to/from wastewater treatments plants in Sweden, see: https://www.ivl.se/download/18.343dc99d14e8bb0f58b51a3/1443173318867/B2208+Mikroskopiska+skr%C3%A4ppartiklar+i+vatten+fr%C3%A5n+avloppsreningsverk.pdf

[22] For further reading on wastewater plants in Stockholm and the membrane filter (in Swedish), see: https://www.stockholmvattenochavfall.se/globalassets/sfa/pdf/informationsbroschyrer/tekniskt-nedslag-i-stockholms-framtida-avloppsrening.

wastewater streams and to prevent or reduce their emission to the environment" (Ben-David et al. 2023: 2). They argue that there is a lack of suitable solutions, other than highly expensive membrane filters, like the one in Henriksdal. The study argues that other approaches are needed. The mucus extracted from the *Aurelia aurita* is one possible solution, and it would work as a bio-flocculant (coagulant) to effectively remove plastics from wastewater.[23] In their study, this Jellyfish mucus outperformed other sorts of coagulants commonly used in wastewater treatment plants today—hence, if it can be upscaled effectively it holds great potential as a tool to increase efficiency in capturing plastics before it reaches the marine environment. And, since a lot of microplastics reach the oceans and the seas via wastewater treatment plants, it is effective to catch them there. The Jelly-Fix project is thus trying to solve two problems at once: the increasing amounts of plastics and jellyfish in the water, and at the same time use less energy than other techniques.

Jellyfish, Care, and Aquabiopolitics

Jellyfish have existed for over 500 million years and are still thriving today. In fact, they are older than the dinosaurs. There are countless types of jellyfish in the oceans and seas worldwide. Some are tiny, others enormous, some can kill humans by their poison, and others have been named "immortals" as they can rewind their cells to become a younger version of themselves. With the increasing acidification of oceans and seas due to human pollution, the numbers of jellyfish are increasing fast. As stated in a 2020 policy brief from the Baltic Sea Centre at Stockholm University,

> More jellyfish but fewer mussels—the expected spread of ocean acidification can have major effects on species composition in the Baltic Sea. It may also make the water slimier and less attractive for swimming. To protect unique ecosystems and future food production, carbon dioxide emissions must be dramatically reduced and measures must be taken against eutrophication, overfishing and releases of hazardous substances.[24]

[23] Due to the small size of micro and nanoplastics one way to catch them in water is to make them lump together. Once they do, they can be filtered out of the water. This reaction is achieved by using some sort of coagulants—like the one being produced from the *Aurelia aurita* mucus.

[24] https://balticeye.org/en/policy-briefs/emerging-ocean-acidification-threatens-baltic-sea-ecosystems/. This citation was collected from this website in 2020.

The Jelly-Fix project focuses on removing one of the hazardous substances mentioned in the policy brief, the increasing amount of plastics. It does this by utilizing one of the results—the increasing abundance of jellyfish. Human practices have created a scenario beneficial for jellyfish but less attractive and not as good for humans. In the story I tell in this chapter, these ancient other-than-humans are used to solve a problem caused by humans (Wakefield and Braun 2019). Here I wish to shift the discussion toward the complex and far-reaching term "care." Care by and for whom? Are we caring for humans exclusively, or is it the oceans' overall health that is being cared for?

The removal of plastics from the ocean can be seen as human care for water and all that lives within it (although humans are the ones causing the problem to begin with, and are the ones who define plastics in water as out of place). We are facing a situation where humans choose which lifeforms to favor over others.[25] As I note in the Chap. 1, care sometimes entails killing (Puig de la Bellacasa 2017). Thus, John Hartigan Jr. (2017: 217) writes that, "Care operates in two basic modalities: a form of attention and a set of practices—interests and activities." His description of care suits the material in this chapter well. There is a caring attention for jellyfish as a species. There is also a set of practices that care for humans over the jellyfish, which illustrates the ambivalence of care and how care is always done with a purpose (Scaramelli 2021). In Chap. 2, I connected with Leslie Sharp's (2019) work on the care of laboratory animals. However, there is one crucial difference between Sharp's work and the case of the jellyfish: the jellyfish are never meant to survive. They are "harvested" for various human uses: removing plastics, fertilizing the soil, providing food, and becoming cosmetics for human beauty. However, in another way, care is extended to the jellyfish as a species. Their importance for the ocean's health is acknowledged, and precautions are taken not to begin harvesting unless their numbers allow it. Jellyfish are seen as a resource that is valuable in the right amounts. However, when too many, they are considered a nuisance and should be managed.

[25] For another wastewater-related study showing how "Caring for wastewater" is a practice of "choosing what to sustain, and who or what may flourish or wither," see Fenna Smits and Mandy de Wilde (2023: 5). This is a practice of doing politics *through* other-than-humans when their bodies become appropriated or removed to fit a human political agenda (Scaramelli 2021).

Situations when other-than-humans are seen as both a valuable resource and an invasive species, or a nuisance, are numerous in the world. For example, Ståle Knudsen describes the case of "the Rapana sea snail in the Black Sea, which is poised between being a useful economic resource and a threatening alien species" (Richardson and Weszkalnys 2014: 8). An example of when a species becomes too numerous and therefore must be managed (as in killed) to not impact the ecosystem negatively is provided by Wanderer (2020: 64–65) who explains the complicated relationship between researchers and wild mice. In her example from a Mexican island, researchers get attached to the mice while knowing they have to be eradicated to safeguard the island's ecosystem. This management of other-than-humans echoes the Jelly-Fix project: removing microplastics from water by sacrificing jellyfish bodies.

Hartigan (2017: 228) further asks, "What happens when care fails?" This is an interesting thought, although my material does not imply failure of care as much as a renegotiation and prioritization of care. Natasha Myers's (2019) description of how life is managed within an enclosure can make explicit how moral orders are staged is helpful here. She illuminates that relations within an enclosed area (in this book, the Baltic Sea and Lake Mälaren) make explicit who is more "important" in a sort of biopolitical ranking of life. Thinking with Myers, we can see that the situation in which jellyfish are used to fix an anthropogenic problem is not one where humans have strategically planned for this situation in the oceans and now harvest the fruits of their plan. Instead, it is an instance of slow violence where humans have turned a blind eye to what the waste they have created has done to the environment around them. Now, when these effects have become too acute to ignore, humans have begun to think about and spend money finding "solutions" to bring the world back to "normal." This is done by controlling certain forms of life, the basis of biopolitics, and by extension aquabiopolitics (Hetherington 2020: 13; Wanderer 2020). Adam (1998: 41) eloquently explains the process leading up to this situation:

> Similarly consequential is the assumption of reversibility, with respect to environmental hazards and degradation it entails the belief that mistakes can be undone, that increased knowledge and better technology can put right in the future mistakes of the past and damage inflicted on the environment now through pollution and the degradation of non-renewable resources.

To use technology to fix a human-made problem is not "staying with the trouble" as Haraway (2016) asks us to do.[26] Instead, I view it as an attempt to rewind history into a time where plastics were not everywhere to be found and when jellyfish did not swarm the oceans and seas as they do today. The partial solution by the Jelly-Fix project suggests transforming jellyfish into biocapital, a resource that can be harvested and utilized (Helmreich 2009). Thus, some types of life are prioritized over others, and actions are taken to enact this priority in an aquabiopolitics of life and death.

References

Adam, B. 1998. *Timescapes of Modernity. The Environment & Invisible Hazards*. London and New York: Routledge.

Ahmann, C. 2018. It's Exhausting to Create an Event Out of Nothing: Slow Violence and the Manipulation of Time. *Cultural Anthropology* 33 (1): 142–171.

Amaral-Zettler, L., E. Zettler, and T. Mincer. 2020. Ecology of the Plastisphere. *Nature Reviews Microbiology* 18: 139–151.

Ben-David, E.A., M. Habibi, E. Hadad, M. Sammar, D.L. Angel, H. Dror, H. Lahovitski, A.M. Booth, and I. Sabbah. 2023. Mechanism of Nanoplastics Capture by Jellyfish Mucin and Its Potential as a Sustainable Water Treatment Technology. *Science of the Total Environment* 869: 171824.

Berwald, J. 2017. *Spineless: The Science of Jellyfish and the Art of Growing a Backbone*. New York: Riverhead Books.

Booth, A., and L. Sørensen. 2022. Microplastics Fate and Impacts in the Environment. In *Handbook of Microplastics in the Environment*, ed. T. Rocha-Santos, M. Costa, and C. Mouneyrac. New York: Springer.

Bradshaw, A. 2022. Can Microbes Be Active Participants in Research? Developing a Methodology for Collaborating with Plastic-Eating Microbes. *Environmental Humanities* 14 (2): 284–302.

Brakstad, O., D. Altin, E. Davies, M. Aas, and T. Nordtug. 2020. Interaction between Microalgae, Marine Snow and Anionic Polyacrylamide APAM at Marine Conditions. *Science of the Total Environment* 705: Article: 135950.

Capolupo, M., L. Sørensen, K. Don Ranil Jayasena, A. Booth, and E. Fabbi. 2020. Chemical Composition and Ecotoxicity of Plastic and Car Tire Rubber Leachates to Aquatic Organisms. *Water Research* 169: 1.

[26] In her book *Staying with the Trouble* (2016), Donna Haraway lays out a competing picture from that of the so-called Anthropocene, one she calls the Chthulucene. Within this story of hers, she suggests that we have to learn how to live and die together on this damaged planet—live and die mutually with other-than-humans and humans alike.

Carrington, D. 2020. Microplastics Revealed in the Placentas of Unborn Babies. *The Guardian*. https://www.theguardian.com/environment/2020/dec/22/microplastics-revealed-in-placentas-unborn-babies. Accessed 12 January 2021.

Davies, E., O. Brandvik, F. Leirvik, and R. Nepstad. 2017. The Use of Wide-Band Transmittance Imaging to Size and Classify Suspended Particulate Matter in Seawater. *Marine Pollution Bulletin* 115 (1–2): 105–114.

de Wolff, K. 2017. Plastic Naturecultures: Multispecies Ethnography and the Dangers of Separating Living from Nonliving Bodies. *Body & Society* 23 (3): 23–47.

Geyer, R., J. Jambeck, and K. Law. 2017. Production, Use, and Fate of All Plastics Ever Made. *Science Advances* 3 (7): e1700782.

Hägg, F., D. Herzke, V.A. Nikiforov, A.M. Booth, K.H. Sperre, L. Sørensen, M.E. Creese, and C. Halsband. 2023. Ingestion of Car Tire Crumb Rubber and Uptake of Associated Chemicals by Lumpfish (*Cyclopterus lumpus*). *Frontiers in Environmental Science* 11: 1219248. https://doi.org/10.3389/fenvs.2023.1219248.

Halsband, C., L. Sørensen, A. Booth, and D. Herzke. 2020. Car Tire Crumb Rubber: Does Leaching Produce a Toxic Chemical Cocktail in Coastal Marine Systems? *Frontiers in Environmental Science* 8: 125.

Haraway, D. 2016. *Staying with the Trouble. Making Kin in the Chthulucene*. Durham: Duke University Press.

Hartigan, J. 2017. *Care of the Species. Races of Corn and the Science of Plant Biodiversity*. Minneapolis and London: University of Minnesota Press.

Helmreich, S. 2009. *Alien Oceans. Anthropological Voyages in Microbial Seas*. Berkeley: University of California Press.

Hetherington, K. 2020. *The Government of Beans. Regulating Life in the Age of Monocrops*. Durham: Duke University Press.

Ialenti, V. 2020. *Deep Time Reckoning. How Future Thinking Can Help Earth Now*. Cambridge, MA and London: The MIT Press.

Liboiron, M. 2021. *Pollution Is Colonialism*. Durham and London: Duke University Press.

McKay, D., E. Stanes, N. Githua, X. Lei, and S. Dixon. 2020. On Global Plasticity: Framing the Global Through Affective Materiality. *New Global Studies* 14 (3): 307–326.

Moeller, S. 1999. *Compassion Fatigue*. New York: Routledge.

Moran, E. 2016. *People and Nature: An Introduction to Human Ecological Relations*. Oxford: John Wiley & Sons, Incorporated.

Myers, N. 2019. Five. From Edenic Apocalypse to Gardens against Eden: Plants and People in and after the Anthropocene. In *Infrastructure, Environment, and Life in the Anthropocene*, ed. K. Hetherington. Durham: Duke University Press.

Nixon, R. 2011. *Slow Violence and the Environmentalism of the Poor*. Cambridge: Harvard University Press.

Pathak, G. 2020. 'Plastic Pollution' and Plastics as Pollution in Mumbai, India. *Ethnos. Journal of Anthropology* 88 (1): 167–186.

Plastic Europe. 2016. Plastics—The Facts 2016. An Analysis of European Plastics Production, Demand and Waste Data. *Plastic Europe*. https://www.plasticseurope.org/application/files/4315/1310/4805/plastic-the-fact-2016.pdf. Accessed 2 March 2021.

Puig de la Bellacasa, M. 2017. *Matters of Care. Speculative Ethics in More Than Human Worlds*. Minneapolis, London: University of Minnesota Press.

Ragusa, A., A. Svelato, C. Santacroce, P. Catalano, V. Notarstefano, O. Carnevali, F. Papa, M.C.A. Rongioletti, F. Baiocco, S. Draghi, E. D'Amore, D. Rinaldo, M. Matta, and E. Giorgini. 2021. Plasticenta: First Evidence of Microplastics in Human Placenta. *Environment International* 146: 106274.

Reno, J. 2016. *Waste Away. Working and Living with a North American Landfill*. Oakland, CA: University of California Press.

Richardson, T., and G. Weszkalnys. 2014. Introduction: Resource Materialities. *Anthropological Quarterly* 87 (1): 5–30.

Scaramelli, C. 2021. *How to Make a Wetland. Water and Moral Ecology in Turkey*. Stanford, CA: Stanford University Press.

Schmidt, J. 2019. *Water. Abundance, Scarcity, and Security in the Age of Humanity*. New York: New York University Press.

Sharp, L. 2019. *Animal Ethos. The Morality of Human-Animal Encounters in Experimental Lab Science*. Oakland, CA: University of California Press.

Smits, F., and M. de Wilde. 2023. "We Are Part of Nature": Caring for Wastewater in an Infrastructural Experiment in the Flevopolder. *Ethnos. Journal of Anthropology* Published Online: 02 May 2023: https://doi.org/10.1080/00141844.2023.2206979

Sørensen, L., T. Gomes, A. Igartua, I. Lyngstad, A. Almeida, M. Wagner, and A. Booth. 2023. Organic Chemicals Associated with Rubber are More Toxic to Marine Algae and Bacteria Than Those of Thermoplastics. *Journal of Hazardous Materials* 458 (September 15): 131810.

Statista. 2021. Global Plastic Production from 1950 to 2018 (in Million Metric Tons). *Statista*. https://www.statista.com/statistics/282732/global-production-of-plastics-since-1950/. Accessed 2 March 2021.

Sun, X., X. Wang, A.M. Booth, L. Zhu, Q. Sui, B. Chen, K. Qu, and B. Xia. 2023. New Insights into the Impact of Polystyrene Micro/Nanoplastics on the Nutritional Quality of Marine Jacopever (*Sebastes schlegelii*). *Science of the Total Environment* 903: 166560. https://doi.org/10.1016/j.scitotenv.2023.166560.

Sun, Y., X. Zhao, S. Qi, X. Sun, L. Zhu, A.M. Booth, B. Chen, K. Qu, and B. Xia. 2024. Polystyrene Nanoplastics Affected the Nutritional Quality of *Chlamys farreri* through Disturbing the Function of Gills and Physiological Metabolism: Comparison with Microplastics. *Science of the Total Environment* 910: 168457. https://doi.org/10.1016/j.scitotenv.2023.168457.

Tian, Z., H. Zhao, K. Peter, M. Gonzalez, J. Wetzel, C. Wu, X. Hu, J. Prat, E. Mudrock, R. Hettinger, A. Cortina, R. Biswas, F. Kock, R. Soong, A. Jenne, B. Du, F. Hou, H. He, R. Lundeen, A. Gilbreath, R. Sutton, N. Scholz, J. Davis, M. Dodd, A. Simpson, J. McIntyre, and E. Kolodziej. 2021. A Ubiquitous Tire Rubber-Derived Chemical Induces Acute Mortality in Coho Salmon. *Science* 371 (6525): 185–189.

Wakefield, S., and B. Braun. 2019. Oystertecture: Infrastructure, Profination, and the Sacred Figure of the Human. In *Infrastructure, Environment, and Life in the Anthropocene*, ed. K. Hetherington. Durham: Duke University Press.

Wanderer, E. 2020. *The Life of a Pest. An Ethnography of Biological Invasion in Mexico*. Oakland, CA: University of California Press.

CHAPTER 5

Interlude: Connecting the Parts

In Part I we have seen examples of aquabiopolitical relations where the current and past human treatment of water has created a situation that not only results in species death in water and bottoms, but also brings about harmful effects that have started to become visible on land. While accompanying marine and plastics scientists, I have shown ways that humans can learn about the marine environment and seen evidence of how human malpractices materialize in water and sediment. The results are anoxic bottoms and escalating methane production that contribute to rising temperatures in the world, not to mention the plastics that swarm in the waters across the globe. However, despite this evidence, not much has changed regarding how humans act toward and with water. One does not have to look further than the 2015 Paris Agreement, or the UN Sustainable Development Goals for 2030, and what has been done since, to reach this conclusion. This is, as I have argued, partly due to an increasing distrust among the public for so-called scientific expert knowledge (Ialenti 2020; Hylland Eriksen and Schober 2017). The situation is evident if one acknowledges the well-known Swedish environmental activist Greta Thunberg's claim that politicians must start to listen to scientists, arguing that they present all the facts while the politicians do nothing. As Thomas Hylland Eriksen and Elisabeth Schober (2017: 7) write, "In this world, where the authority of science and empirical methods is being questioned

R. Rodineliussen, *Underwater Worlds*, Sustainable Development Goals Series, https://doi.org/10.1007/978-3-031-63370-6_5

and where even world leaders may brush aside uncomfortable facts as 'fake news,' it is increasingly difficult to know whose knowledge to trust." Schober (2017) uses an example from the Philippines to exemplify this line of reasoning, explaining how the country is caught in a double bind between energy and climate: rising temperatures, a result of global warming, increase the demand for energy. And due to the proneness of the Philippines to experience environmental disasters, many argue that coal mines are the most reliable source of energy for the future. However, the use of coal contributes heavily to climate change and exacerbates the problem. Economy, energy, and ecological survival are here intertwined in a nexus of relations that are extremely complicated to navigate, making it difficult for people to know what knowledge to trust. Another recent example of how some politicians ignore, twist, or brush aside scientific knowledge can be seen in Sweden. In a support party to the newly elected government from 2022, one of the leading politicians recently declared the climate crisis a hoax that lacks scientific credibility.[1] This statement is in direct opposition to Thunberg's claims, and also to what we have seen in Part I. In addition, the new government of Sweden has also discontinued the Environmental Department that has existed since 1987.[2] Hence, facts alone are apparently not enough to cause people to take action for the environment.

In Part II, we will explore what Hylland Eriksen (2016, 2023) has called attention to: that in our current "overheated" world, scientists, environmental activists, as well as anthropologists must acknowledge and address the natural and cultural worlds in a way that highlights their interconnectedness and interdependence. If this message is not understood by authorities and the general public we are heading toward a dangerous future. Our focus will be on an initiative of trash scuba divers who confront these facts in a hands-on action-oriented manner. This group of trash scuba divers seeks to bring the public and authorities into a direct relationship with, and experience of, the waste leaked from human societies into the water. Their objective is to convince people living on the surface of the need to act now, while providing one example of how this acting can be done. The divers believe that if people get to know the harm

[1] https://www.aktuellhallbarhet.se/miljo/klimat/sd-politiker-klimatkrisen-fejk%2D%2Dsaknar-vetenskapligt-stod/ . Accessed April 21, 2023.

[2] https://www.aktuellhallbarhet.se/miljo/miljopolitik/efter-35-ar%2D%2Dnu-laggs-miljodepartementet-ner/. Accessed April 21, 2023.

the waste can do to the marine environment by other means than just scientific facts, it will improve the chances that individuals will begin to care enough to change the way they treat waste and water. To achieve their goal, the group builds installations made out of waste and invites the residents of Stockholm to engage with it in different ways, utilizing senses other than vision, such as touch and smell.

Parts I and II differ in their ethnographic focus. The first part looks at marine scientists, and the second at environmental activism. Moreover, the parts are, in a way, partly focusing on different knowledge regimes: the cognitive and the embodied. The contrast between knowledge regimes has a lengthy history within the discipline of social anthropology, and may point to important relationships with the social world (Hylland Eriksen and Schober 2017: 9). Part I focus on fact based expert scientific knowledge. Part II is about hands-on, sensorial, and embodied experience. Analytically I follow the same trajectory. While my larger theoretical frame is the anthropology of the environment, water, and waste, as well as that of political ecology, in Part I I hold hands with STS scholars, and in Part II I work with those who pursue an embodied sensorial experience of the world. Meanwhile, I remain conscious of the other-than-human presence and role in the world. This eclectic theoretical perspective fits the material I present and the conclusions I draw from the ethnography. STS provides tools to engage with scientific practices, and the sensorial-embodied with ways of engaging the underwater world through bodily immersion and physical contact. The other-than-humans are ever present in the environment under study, either through scientific instruments or scuba diving.

My objective is to show that although "on land" and "underwater" might appear to be two separate worlds, they are closely interlinked and dependent upon one another (Bubandt 2023: 185; Pauwelussen and Verschoor 2017). It is of uttermost importance to follow the relations that link these worlds together if we are to understand them and find possibilities for change (Strang 2023). Or, to use Hylland Eriksen's formulation, "It is important to follow these relations if we hope to escape the double-bind of modernity where culture has damaged nature to such a degree that nature now needs the help of culture to protect it against culture itself."[3]

[3] From a conversation with Thomas Hylland Eriksen.

Throughout the **book**, I show how human societies exert power over living marine organisms in a desire to make life on land as pleasurable as possible for humans—what I call aquabiopolitical relations. In pursuit of this understanding, we trace the relations, life, and materia that make up the assemblage of water in the Baltic Sea and Lake Mälaren. It is important to understand this human-altered assemblage of water in order to grasp the scale and volume of human power over water and marine organisms. Small changes in how humans act toward water can have huge consequences for life in the water, and by extension, for life on land. Levels of nutrients, amounts of plastics, and even quantities of heavy metals such as lead, are all part of the water assemblage, and these levels mainly depend on human actions or inactions.

This discussion is ethnographically grounded in the context of Stockholm, a city proud of being environmentally friendly and an international role model in the battle against climate change. I show how scientists and divers, in different but aligned ways, have made the case that the authorities of Stockholm—and Sweden at large—are not as environmentally oriented as they wish to appear. Moreover, the water around Stockholm that is "known" to be clean, has been shown to contain waste and pollutants in large quantities and shapes, to the degree that one can almost claim the water is unclean. That people can drink tap water in Stockholm is a well-known "fact." However, the material I present in Part II does, in a way, question this assumption. Let us investigate further by moving on to Part II and the trash scuba divers.

References

Bubandt, N. 2023. Tidalectic Ethnography. Snorkeling the Coral Reefs of the Anthropocene. In *Rubber Boots Methods for the Anthropocene. Doing Fieldwork in Multispecies Worlds.* Minneapolis: University of Minnesota Press.

Hylland Eriksen, T. 2016. *Overheating: An Anthropology of Accelerated Change.* London: Pluto.

———. 2023. Threats to Diversity in the Shadow of Anthropocene Overheating: A Biosemiotics Perspective. *Kritisk etnografi—Swedish Journal of Anthropology* 6 (1): 9–22.

Hylland Eriksen, T., and E. Schober. 2017. 1. Introduction: Knowledge Regimes in an Overheated World. In *Knowledge and Power in an Overheated World*, ed. T. Hylland Eriksen and E. Schober. Department of Social Anthropology: University of Oslo.

Ialenti, V. 2020. *Deep Time Reckoning. How Future Thinking Can Help Earth Now*. Cambridge, MA and London: The MIT Press.

Pauwelussen, A., and G. Verschoor. 2017. Amphibious Encounters: Coral and People in Conservation Outreach in Indonesia. *Engaging Science, Technology and Society* 3: 292–314.

Schober, E. 2017. 3. The Frailty of Power. Electricity Generation and Sustainable Livelihoods in Zambales, Philippines. In *Knowledge and Power in an Overheated World*, ed. T. Hylland Eriksen and E. Schober. Department of Social Anthropology: University of Oslo.

Strang, V. 2023. Relating to the River: New Bio-cultural Diversities in Human Engagements with Water. *Kritisk etnografi—Swedish Journal of Anthropology* 5 (1): 33–47.

PART II

The Trash Divers

CHAPTER 6

No One's Water: Trash Scuba Diving in Stockholm

In this chapter, we will get to know the members of Rena Mälaren while learning what it is to "trash dive" and the technical equipment required to do this work. We will also look at the role of the "rope puller," the person hauling up the underwater trash found by the divers. The chapter will further explore the relations between the divers and the authorities of Stockholm as well as with the general public, and, ultimately, the marine environment. We will seek to understand the aquabiopolitical relations at play, and what the trash scuba divers hope to achieve with their actions. The chapter begins by outlining how the trash scuba diving group Rena Mälaren came to be. We then meet a few of the members and look more closely at the role of a trash scuba diver and rope puller respectively, before moving on to mapping and thinking about the relationship between humans, waste, and water in the context of Stockholm. Along the way, we will see how the group of volunteers grew, while also recognizing the impact of the Covid-19 pandemic.

The Origin of Rena Mälaren

My first conversation with Fredrik Johansson, the founder of Rena Mälaren, took place on a windy September morning in 2019. At this time, he and a friend were the only ones conducting trash dives together. When I asked Fredrik how he got the idea to begin this work he replied:

R. Rodineliussen, *Underwater Worlds*, Sustainable Development Goals Series, https://doi.org/10.1007/978-3-031-63370-6_6

> I got the idea early summer of 2019 when I helped John to dive below his boat. I was going to have a look so that all was fine and check if something needed to be fixed. But when I was down there, below the boat, I noticed something that looked like a car battery. I did not believe my eyes and had to get closer to check it out. Indeed, it was a car battery. Then I looked around further, and to my dismay, I saw battery after battery along the bottom. I returned to the surface and asked John for a rope to help me pull the batteries up out of the water. Well, we pulled batteries all day. In the evening, we discussed this and decided that the two of us would see how many batteries we could pull out of the water during the summer. As the summer went on, the number of batteries surpassed my expectations, and I understood that to really make a difference, I needed more help. Therefore, I have now reached out via local media and posted an advertisement on the Swedish Dive Society's website, dykarna.nu, to find more divers and attract volunteers who can help me in this endeavor. This was the way you found out about me too, right?

In the beginning of the diving project there were just a handful of individuals involved. They were a mix of newcomers who found their way to the dives through Fredrik's advertisement or via small articles about the project in local newspapers, and some were Fredrik's colleagues from his time as a diver in the army. At first these individuals came together on Saturdays to conduct trash dives, but one day a week quickly turned into two, or even three. However, in Stockholm (as in most urban areas) you need a permit to scuba dive. These permits normally have to be submitted on paper before the planned dive, and the divers must keep the permit with them at the dive site, should the authorities come by to check. But early on, Fredrik developed a trusting relationship with Stockholm Harbors, those responsible for the permits, which meant that he was allowed to obtain permits by phone—he just had to call and inform Stockholm Harbors prior to a dive, telling them the approximate dive time and where the dive would take place. This only applied to "ordinary" areas; if the dive was planned close to, say, the Royal Palace, other types of permits were needed. Fredrik explained, "Stockholm Harbors allowed me this convenient solution because they like the work I try to do and want to assist with making it less complicated and time consuming to dive regularly." (Fig. 6.1).

Another logistical aspect of trash diving that had to be solved was how to dispose of the collected waste. On a few occasions in the beginning, more regularly later on, Stockholm Harbors also assisted with this by

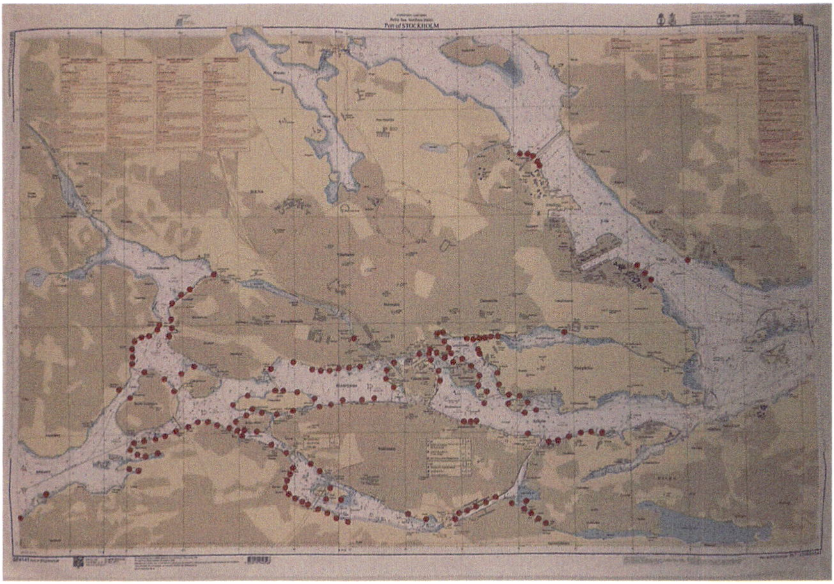

Fig. 6.1 Sea chart of central Stockholm showing the central areas in which Rena Mälaren has dived between autumn 2019 and autumn 2021. Many red dots indicate more frequent diving in the area, rather than a tally of the actual number of dives. This chart was used by the author to document how the diving project developed during fieldwork. (Photo: Author)

providing large containers that the divers could fill with waste. Employees at the waste management stations later sorted the waste. However, during the first year of the project, until mid-2020, the divers had to dispose of the waste themselves on most dives. They did this by driving it to waste management stations. But, as private persons, there are regulations as to how much waste one can leave at any one time. For example, you can only leave four car tires on one visit. This is a problem when, on an ordinary dive, the group collects between 10 and 40 tires. On these occasions, they have been forced to drive from one waste management station to the next until they have disposed of all the tires. As Fredrik told a new diver who inquired about how the waste was disposed of after a dive (Fig. 6.2),

> It is just crazy! I mean, they are not even our tires. We bring them up after others. They are in the water, leaking toxicants and polluting. Then when

Fig. 6.2 Left: Fredrik's car full of car batteries and other waste; Right: Car tires collected during one dive at Söder Mälarstrand. (Photo: Author)

> we remove them, we are not even allowed to dispose of them at the waste management station. I have been in so many arguments with the employees about this, but they are stubborn in following the four tires-rule. I have called management and even the city department responsible for the stations, but no, we are still, most of the time, not allowed to dispose of more than four tires at a time.

The reason for the lack of response from the authorities regarding the agency that should take care of this waste from the water is, according to them, that there is no consensus on who is responsible for waste below the water. On land, there are clear borders between different city areas, but this is not the case with water. As one of the divers stated, "It seems like water is a no one's area, at least when it comes to the responsibility to keep it clean." To make the cleaning of water someone's responsibility was one of Fredrik's initial goals. His intention was not to force them into cleaning, but to know whom to speak with when it came to the disposal of the waste that Rena Mälaren brought up. In the late summer of 2020, the city took this responsibility upon itself, as we shall see below.

Several scholars, including Rademacher (2015), Drew (2020), and Cornea et al. (2016), have pointed toward the need for an urban political ecology focusing on water to address the issues of environmental politics in relation to economic forces in an urban setting. For example, Karen Bakker (2010: 7) notes that it is important to follow how water "moves through the city" to understand how these relationships are created. Relations become visible when one looks at both the movement of water and the items and materia in the water. These relationships are what matter, should one want to change the way the bottoms around Stockholm have become. If focusing purely on the sources of waste or the end destination—the bottoms—one misses the full picture. The divers are engaged in mapping these relations, trying to figure out who is responsible for each step of the process, from when and how waste "leaks" out of the waste management system (Reno 2016) to how the waste then travels and ends up in the waterways. Who is responsible and at what point, what are the relationships between these roles/persons, and how could these relationships be altered to change the outcome, are some of the questions the divers from Rena Mälaren ask the authorities. In my analytical work, I pursue the same set of questions as the divers. I also consider the diver's role in engaging with these questions in a relationship with the public and the authorities. However, let us now engage with some of the members from the group before we look at what it entails to be a trash scuba diver and a rope puller.

The Members

Fredrik told me that from day one, the idea behind Rena Mälaren had been to be inclusive and welcoming to all who share the aspiration of keeping water free from waste. Following this aim, the group has grown to include volunteers from all parts of society, from leading business people and famous adventurers to students, pensioners, the unemployed, and immigrants. The group is as varied in economic class (high to low) as it is in gender (the male/female ratio is about 50%), and in age (5–80 years). They welcome whole families to join in on cleanups, where everyone can assist to the best of their abilities. Children, for example, can walk around and collect waste on land near the dive site, thus actively contributing to the cleanup work. This mix of individuals from all strands of life can seem chaotic, but according to members and those joining for cleanups, these connections between people are what makes Rena Mälaren special. One diver in his early 40s whom I will call Per[1] put it this way,

[1] All divers and rope pullers except Fredrik will be given pseudonyms.

> We are a group of distinct personalities—no one is like the other, but this is also what makes us into such a lovely group to be part of. Since no one is similar to the other, but shares the joint interest to clean our waters, no one is also different. We are good with and for one another. We are like one big family.

As I will elaborate below, the group's work consists of two roles, that of "trash diver" and "rope puller." These roles, however, may be combined. Some members move between them, acting as a diver on some cleanups and as a rope puller on others. This is why I place emphasis on the *roles* more than on individuals. However, an approximate description of the people who mainly occupy each role (with the flexibility mentioned) is appropriate. The divers are between 18 and 55 years old, with a slightly higher percentage of men over women. They are also mostly from the middle-class or higher, due to the high cost of buying diving equipment (I return to the financial aspects of diving below). The rope pullers are of all ages and economic classes, divided equally between males and females, and include pensioners.[2]

The pensioners are a category of their own. As they have a lot of free time, they are one of the most active groups. Take Olle, for example—he is a man in his early 60s on early pension (he could finish work early due to a deal with his employer). For this reason, he got a lot of time and energy to spend on interests of his—he decided to spend the time on Rena Mälaren. When asked about his choice he responded like this,

> I had just left work and started to think about what I should do with my time, now that I am no longer going to work. I am not one of those people who can just sit at home. No, I would go crazy. So, I was looking around for interesting projects. At this time, I noticed a small newspaper article about Rena Mälaren. It included the contact information to Fredrik. Hence, I called him the same day and asked how I could join in the work. Fredrik asked me to just come down to one of their next cleanups—they had one already the next day. I went there, and have not stopped coming since.

[2] The well-known Rotary Club frequently collaborates with the divers as rope pullers. This high-end environmentally engaged club works closely with Rena Mälaren. More information about Rotary can be found here: https://www.rotary.se/ as I will not elaborate more on this organization. When they take part in cleanups they work just as other rope pullers, and will therefore be treated in the same way in this book.

People like Olle have become the backbone of the organization, due to both their enthusiasm and their ability to take part in dives on any day of the week. Some of the divers, like Fredrik, have managed to make special arrangements with their employers so that they can conduct trash dives on work days, as long as the workplace is mentioned now and then when results from these dives are made public. This is one way these workplaces support the environmental work. Fredrik's employer, for example, is one that provides meaningful activities for individuals who for various reasons might have trouble holding a regular job at this time. Therefore, on what Rena Mälaren has come to call, Monday Dives, these individuals join in and act as rope pullers together with other volunteers. I will not be naming any of these individuals or describe them further for ethical reasons, but I will mention that if they can choose between activities, they oftentimes chose to join the option with the dive together with Fredrik. There are also a number of students from Stockholm University's Marine Biology Department who share their increasing knowledge about life beneath the water with others in Rena Mälaren. This is especially valuable for children who can learn what this fish or that "animal" is that came up inside a car tire, and find out about the possible harm done to them from toxicants leaking from the waste.

To emphasize the great variety of personalities in the group, there is one young diver in her early 20s, Camilla, who, although being a scuba diver, is afraid of fish. "I would never dive in tropical waters with all that fish. But here, in Lake Mälaren, I do not see many fish so it is okay." Camilla further elaborated on the feeling of diving, making the case that it is the embodied presence of her body in the water that makes it so attractive to her:

> I love the feeling of weightlessness, how I can move around freely without touching anything but water. The darkness is, to me, meditative. I start to feel my body more as I move around maneuvering with the help of touch rather than that of sight.

The importance of the meditative aspect of diving is acknowledged by most of the divers, and is one of the reasons they give for not being afraid of diving in water without visibility. Yet, as Per stresses, "You need to be a bit special, this is not for everyone. We are all crazy, but good crazy!"

There are also considerable differences among the divers when it comes to diving background. Some, like Fredrik, got their experience from diving for the military, searching for explosive mines in the water.[3] Others, like Per, started out as recreational divers on holiday in the tropics, but have decided to put their diving knowledge to practical, rather than recreational, use. A few, like Evelyn (in her early 40s), work as full-time diving instructors. Still others, like Camilla, obtained a diving certificate with the specific aim of diving for trash with Rena Mälaren. Because skills and backgrounds differ among the members, much attention is given to practicing safety routines to ensure that no accidents will occur. Instructors like Evelyn can be of assistance with this by organizing training sessions that combine underwater movements for divers with exercises on how rope pullers on land should assist divers in case of a diving accident. The group conducts a minimum of two training events annually to make sure they do all they can to prevent accidents.

Trash Scuba Divers

A trash scuba diver is a denomination for an "ordinary" diver who dives after trash in the water. There are worldwide organizations conducting trash dives, such as *Project AWARE,* that will be discussed in Chap. 8, but most of these work in open water where there is good visibility. What makes the trash dives carried out by Rena Mälaren extraordinary is that they are done in an urban center and in water with extremely low visibility, and sometimes none at all. There are many additional dangers when diving in urban areas in comparison to ordinary recreational diving—most of all, the boat traffic. A boat can be a lethal danger to divers beneath the surface, especially low-lying ferries in shallow waters that can easily ram a diver should they come too close. There are also many buoys (floating devices used for tying boats that are anchored to the bottom with metal chains), lines, cables, and a variety of waste in urban waters that a diver can get caught on or hurt by, should the object be sharp. An experienced diver could easily avoid these objects if there is enough visibility to detect them in time. However, in Lake Mälaren (and other waters in urban areas)

[3] In the military divers are trained from "scratch." The army does not allow you to become a diver if you have earlier dive experience. They want you to be without prior knowledge in order to mold you into their type of diver. This differs from the trash scuba diving group, who want a new diver to be experienced, to have good buoyancy, and knowledge about their equipment. But then they relearn how to dive, as trash diving differs greatly from recreational diving, as will be explained in this chapter.

visibility rarely extend beyond two meters, in fact often less than one meter, and even down to 10–20 centimeters. Moreover, as soon as the bottom is touched, sediment rises into the water, effectively removing any visibility entirely. The particles disturbed on the bottom are then slow to settle back down, so visibility will be impeded for some time, depending on currents.

A trash diver in Stockholm must learn to safely move around along bottoms, below boats, and among waste and cables, without getting caught or hurt. When below water without visibility, it is absolutely essential that a diver can remain calm, in a way, emptying oneself of feelings that could trigger panic, and thinking calmly about how to move about (Merchant 2011: 54; Ecott 2001). Panic is one common reason for diving accidents. When a diver panics, the diver will no longer think logically, and that is when accidents happen. Diving as a practice is full of rules that cannot be disregarded, such as how quickly you can resurface from a certain depth, and if you need time close to the surface to reset the pressure within your body before exiting the water. To avoid many of these possible dangers, Rena Mälaren only conducts trash dives in water with a depth of less than 10 meters. When a dive is shallower than this, the body will not get hurt by the changing pressure, should the diver exit the water too quickly. This is one way to make trash diving safer. Below water, when you cannot see, directions are hard to keep in mind, and becoming lost is easy. You might think that you are on the way back to land, but actually, you are heading in the opposite direction. Here the rope connecting the diver with the rope puller on land is critical, as will be discussed below (Figs. 6.3–6.4).

In the opening vignette of the Introduction, I described diving without visibility and the way a diver must sensuously feel one's way around under water. But in order to be able to dive that way, totally engaging one's senses to experience the underwater world, a diver must have embodied skills of how to manipulate the equipment (without being able to see it), and preferably without even thinking—it should just happen. Divers with considerable experience below water know their equipment by heart so that it often becomes like a second skin. They learn how to sense even the slightest pressure on one of the fins or to gently feel what type of surface they are touching, even through 7-millimeter-thick three-finger gloves, with fingers stiff from the cold. A novice to diving will feel clumsy in all the gear needed for cold water diving, but after many hours of practice the suit stops being cumbersome and starts to be easy to move in as long as you are in the water. In the methods section of the Introduction I develop

Figs. 6.3–6.4 Trash scuba divers entering the water. (Photo: Joakim Hall)

this embodied/emplaced and sensorial experience of diving, especially with regard to cold, low-visibility water. However, one aspect that needs some further elaboration is "touch." Hayward (2010: 582) writes that "embodied vision is necessarily accreted by the other senses and their amplification." Trash scuba divers do indeed "see," but they do so without sight. It is the other bodily senses that are attuned to feel/touch the underwater world in such a way that the diver can tell the surroundings without any visibility. Furthermore, Puig de la Bellacasa (2017: 96) writes, "Haptic engagement conveys an encouragement for knowledge and action to be crafted *in touch*." This knowledge-action trough touch applies to both the divers below and the public above water in how they learn about underwater worlds. Touch as a sensorial way of knowing and relating is one further step away from the cognitive, and toward an embodied way of knowing the world, as discussed in the Interlude. This perspective might assist to make knowing (but inactive) individuals on land begin taking action to assist the world below water. For example, it is hard to touch the slimy surface of a car tire brought out of the water without also feeling touched by it, disgusted, and unwilling to get it close. Hence, I understand touch to have an impact, and that it is a way of knowing that might kick-start individuals into action. Let me exemplify this sensorial experience of knowing the water and objects within it through touch with a dive in the central area of Norrtälje, a small city just outside of Stockholm. The visibility here was absolutely zero, just like the dive in Västerås described in the Introduction.

As I submerged into the brownish water and visibility disappeared, I slowly lowered myself until I could touch the bottom with my hand. I was holding a large metal hook in my right hand that was connected to the rope and the rope puller, which I used to attach to trash objects in the water. This hook I would use onward to navigate the underwater world. As I moved forward in the water, feeling my way with my hands, I used the hook to gently knock against items I encountered. When I did, different types of material produced various sounds and also felt different when knocked on. Metal produced a clear "clonk" and a rigid surface; stone a softer "clonk" and a rougher surface. Wooden objects made almost no sound and felt soft when hit with the hook. Knowing if the object in front of me was made of wood, stone, or metal was very useful in order not to have to go up and down too often only to realize that the rope pullers had pulled wood or stone out of the water. Such material is not doing any harm to the marine environment and, therefore, should be left in place. Since my vision could not assist me in localizing what objects to bring up out of the water, I instead utilized sound and touch to sensuously allocate and differentiate between what to bring up and what to leave in the water. Moreover, the rope that was held slightly tense by the rope puller allowed me to know in what direction there was land, and therefore in what direction to move. If I gently pulled on the rope, I could feel the direction it came from and knew where to return. Trash scuba divers learn to adapt to and make use of various senses and objects in similar situations. Now, let me elaborate on the role of technology for a diver.

Is a Scuba Diver a Cyborg?

A cyborg is a creature that is not only human, and not only a machine, but a human and machine that live and act together (Haraway 1991: 149). Earlier, cyborgs were discussed in relation to the research vessel *Electra* and compared to Helmreich's (2007) account of the submersible named *Alvin*. Yet, as I argue, the relationship between human and technology in the case of *Electra* did not fulfill the criteria of a cyborg. For scuba divers, however, the notion of cyborg appears appropriate. A scuba diver is a human who can be below water for hours, something not natural for humans since we cannot breathe under water. What makes it possible for a scuba diver to stay below water is the scuba technology that functions as mechanical lungs, providing the human body with air. Other equipment provides the body with protection from cold, assists it to keep buoyant, maximizes its ability to move, up and down, side to side, and, not least of all, to look around in the water. All of these things are possible for the human body due to the scuba technology. In this light, a scuba diver can be seen as a cyborg, that is, a manifestation of human and technology working as one (Raycraft 2020; Jue 2020).

When a person studies to become a scuba diver, one of the first things they learn is to breathe through a regulator under water—to trust the technology. It is common to feel awkward in the beginning, to be insecure about the reliability of the regulator to actually provide air when you breathe. The first few times beneath the water, breathing with the aid of the scuba technology, one often feels strange and anxious. When you take a breath, dry air comes into your mouth, followed by a rather loud sound when you exhale into the water, in the form of bubbles moving toward the surface. Below water, sound is very different: you can clearly hear every breath you take. For this reason, many divers describe being below the water as a time when you listen to your own breathing. However, once you become accustomed to this sound and the equipment, other sounds begin to filter through the noise of your breathing. A trash diver must learn to master the technical diving equipment so that there is awareness, sensitivity, and vulnerability toward the surrounding marine environment (Jue 2020: 45). This will enable experiencing and gradually understanding the underwater world: a way to view it from below (Raycraft 2020). Divers who have mastered their equipment become, one might say, underwater cyborgs, since it is advanced technology that allows a scuba diver to stay below the surface. But another technology, a simpler one, is of similar importance to the work of a trash scuba diver—the rope, and the rope puller.

Rope Pullers

The role of the rope puller is crucial in the work of Rena Mälaren. Fredrik often tells the land-crew:

> It is you, on the surface, who do the hard work. You are the ones that actually haul all the trash out of the water and carry it to the containers. We divers could dive all we want, searching for waste, but without you, it would stay in the water.

Pulling the waste from the water and carrying it to the container is heavy work. For example, car batteries weigh 20–80 kilograms, and metal waste can be even heavier. Many items also weight more when they are wet. So, for example, car tires collect mud inside, making them much heavier when pulled out of the water. A regular cleanup session collects between 500 and 3000 kilograms of waste, divided between, at an average, six rope pullers working for two to three hours (Fig. 6.5).

The rope pullers are also essential for keeping the divers safe while they are below water. Each diver is paired with a rope puller; they are then connected by a single rope. The diver is strictly forbidden to let go of this rope while below the surface, and the rope puller must hold the other end of the rope at all times. But the task has more to it than just holding the rope.

Fig. 6.5 A rope puller in action. (Photo: Anton Ivo Martin)

The rope puller must learn to hold the rope tight enough to feel the diver's movements, but not too tight, as that would disturb the diver. There are several reasons for this line between diver and rope puller. The primary one is that should the diver get caught in something below water, he or she can give three long and hard pulls on the rope. This signals to the rope puller that the diver needs assistance, and the rope puller will immediately send another diver to the one who is caught. The other diver can then quickly follow the rope to find the diver having difficulty. Rope pullers also use the rope to signal to the divers below using the same three pulls. This tells the divers to make their way safely to the surface. This signal can be used when a diver is needed to assist another diver, or if, a ferry or a boat is approaching that might become dangerous for an unaware diver beneath the water. Finally, the rope is used by the diver to connect to objects underwater that the rope puller is to pull up. An experienced rope puller will also track the diver's movements via bubbles on the surface that follow the diver. A steady stream of bubbles, together with slight tension on the rope, allows diver and rope puller to be aligned even though they are on different sides of the water's surface.

To exemplify the procedure, I will recount one dive close to the National Museum in Stockholm in the winter of 2019. I acted as a rope puller. The dive began in the parking lot of some large ferries bound for the Stockholm Archipelago. Prior to the dive, inquiries had been sent to the ferry captains. They had promised Rena Mälaren a time window of three hours in which there would be no ferries coming in. We all arrived, the divers geared up, the rope pullers arranged their ropes, we put out our dive flags—an internationally recognized blue-white flag that indicates that there are divers in the water—and everyone paired up. All seemed to be going well. The divers found things and the land crew pulled them up. But then, as one of the divers ventured further away (still connected by the line to his rope puller), I noticed a large ferry coming right at us, and rather quickly. The diver's rope puller gave the signal of three long and hard pulls to the diver to call him back. I and the others started waving and shouting at the captain of the incoming ferry, one of us holding a dive flag in the air. When the ferry was only a few meters away, it stopped. The diver was just resurfacing near the parking lot, but still in danger. A diver cannot bounce right up, even when a sign of danger is given, and it is difficult to know how to act when a ferry is approaching. Usually, one would press oneself against the bottom of the water to gain as much distance from the boat as possible. But here, the ferry was riding so low in the water that it would not have helped: the diver would have gotten crushed between the sea bottom and the ferry. The sound of a ferry below water is also terrifyingly loud, vibrating from everywhere at once, disorientating the diver as to where the ferry is coming from.[4] Thus, the diver has no other option than to follow the rope back and hope that the land crew will make sure the ferry is kept at bay. After the ferry stopped, the captain rushed from the bridge and called to us in a deeply apologetic tone, "I am so sorry, we totally forgot that you were here. Luckily you made me see you in time!" Hence, even when permissions have been obtained before a dive and security measures and warning signs exist, danger can still be present. For this reason, it is essential that the land crew remain alert to what is happening around the dive site.

Among the rope pullers there is always one who has the overall responsibility for the dive and who makes decisions at short notice when needed, the *dive leader* (Fig. 6.6).

Fig. 6.6 Dive leader. (Photo: Oskar Kihlborg)

[4] The human ear is not adapted to the way sound moves below water, and it cannot determine from which direction the sound comes. Therefore, sound below water is felt as if coming from all directions at once. This makes it hard for a diver to know where a boat is approaching from, and in what direction to swim in order to avoid impact.

The dive leader is the communication link between the entire dive and those outside of the given group of divers, such as the police, the public, journalists, and others. The dive leader is also the one you inform if a diver gets stuck, if for some reason you need to leave and your diver is still in the water (another rope puller has to be paired with the diver), or if you need any other sort of assistance. The dive leader always wears a yellow vest with the label *Dykledare* on the back, as seen in the picture above. Who the dive leader is varies between dives, but it is always one of the more seasoned divers or rope pullers, like Per or Olle. It is vital that they be very experienced so that they can make informed decisions. Having a dive leader is essential to assure that all the rules of safety are followed on each dive. For example, the dive leader keeps track of how many divers are in the water at any given time. As soon as one diver exits the water, the dive leader must be informed. In this way, the group guards against the possibility of "forgetting" a diver in the water.

As I show in this section and the one above, both advanced scuba technology, and "simple" technology as in the rope, are vital for the work of trash scuba divers to function. Now, let us continue the story about how Rena Mälaren expanded from a few individuals to a large non-profit organization, and see to their relationship with the authorities and the marine environment in Stockholm.

Expanding

In the autumn of 2019, I conducted a number of interviews with officials from the local government of Stockholm and other environmental organizations in Stockholm such as *Håll Sverige Rent* (the largest cleanup organization in Sweden) and *Städa Sverige* (the Swedish Sport Department's cleaning organization). All had heard about Rena Mälaren but none wanted to be involved with a group of trash divers, as they believed this work to be too dangerous. If an accident should occur, they did not want to be associated with the group. However, the continuous work by Rena Mälaren, with weekly cleanups throughout the year and no accidents being reported, together with increasing public support, made it more and more attractive for the authorities and organizations to begin collaborating with the divers.

Thus, what the authorities and large organizations at first believed to be a short-lived and risky undertaking proved itself to be a growing force with broad public backing. This development had several causes. The divers came up with many ideas that captured the attention of the public, making the group attractive for traditional media to cover. The divers also became active users of social media. The quickly expanding public support made the group interesting to the authorities and large organizations—now not minding the possible risks—as being associated with the group at this time commanded good public attention. Then the global pandemic came along, and the group quickly grew from a handful of individuals to well over 300 active volunteers, divers, and rope pullers. However, scuba diving is not cheap, and trash diving involves considerable wear on gear, so the issue of finance became important to the group.

The Work of Financing Trash Diving

A diver must be equipped with a dry suit to dive in Swedish waters for most of the year. Such a suit is made of composite material or crushed neoprene with built-in boots and sealing at the wrists and neck to keep water out and the diver dry. It allows the diver to wear dry, warm clothes beneath the suit, making it possible to dive in water with temperatures of 2–10 degrees Celsius, which is the general case in Sweden. The diver also needs gloves (dry or wet), a hood, goggles, fins, weights, an oxygen tank, a regulator, a buoyancy control device (BCD), a pressure gauge, a dive computer that shows depth, and a diving flashlight.[5] These are the equipment basics. Then there are other things, such as a dive knife that add safety for the diver. To buy a basic level set of dive equipment is expensive, costing about 40,000 SEK (approximately $4700). The suit can easily tear if it should come in contact with sharp objects, and metal waste in the water is often very sharp. Thus, suits and gloves are routinely damaged, resulting in further costs to the diver. The divers in the group are willing and able to buy their own gear and use it for trash dives, but if they had to pay for repairs as well, they would soon lose interest because it would be too expensive.

[5] The flashlight is important because these waters are very dark. However, using it is not always possible or even advisable. If you dive in shallow water (3–5 meters) it is sometimes better to rely only on the light penetrating the water. This is because during considerable parts of the year there are so many organisms in the water that they reflect the rays from the flashlight. This results in an effect similar as to when you use the headlights on a car in a snowstorm. The light reflects back into your eyes and blinds you, rather than assisting you to see further away.

In addition to the cost of dive gear, the work of Rena Mälaren involves expenses on land. Rope pullers need warm, waterproof clothing that may become extremely soiled. There is a saying in the group, "If you're not dirty, you have not done your job." The rope pullers also need suitable gloves to protect their hands, rubber boots, good quality ropes that will not break, strong metal hooks that can be attached to the ropes so that divers easily can secure them to items along the bottoms, as well as fuel for cars and some food for all to congregate around after the dive.

At first, these costs had to be covered by the individual divers and rope pullers, as the group had no external funding. Then it was realized that by returning car batteries, the group could earn the equivalent of $.40 per kilogram, and other metals could also be sold as scrap. For example, copper returns $3.50 per kilogram. If one only counts the car batteries retrieved, which by mid-2022 totaled 33,000 kilograms (more on this below), this recycling has generated $13,200 plus additional money for metals such as copper—enough to repair some equipment. Thus, the divers could turn some waste into smaller sums of money that could pay for repairs and buy essentials for the cleanups. One significant cost associated with regular diving is refilling the tanks with air. This cost has been eliminated through the help of a dive center that offered the group free refills as a way of supporting the work. This center has also made a deal with the divers that all members get a 10% discount when buying equipment through them, and this has in turn led to many purchases and repairs being made through this center.

In this way, the divers found value in what others had discarded (O'Hare 2019), capitalizing on the creative opportunities available in the management of waste, and creating a sort of emerging waste infrastructure (Reno 2015). Taking car batteries as one example, it is possible to analyze them using Michael Thompson's (2017) "rubbish theory." Thompson argues that objects can be categorized under three heads: *durables* (valuables), *transient* (of degrading value), and *rubbish* (an ambivalent category that can align with the transient or be made into durable once again). A car battery begins as an object of value (durable) when providing energy for a car. Once the energy is depleted, the battery is discarded and seen as transient (waste) and dumped in the water. The divers then bring the batteries out of the water and by reusing the lead inside the battery, *they turn what was waste into something valuable*—the capacity of an object from the rubbish category (Rodineliussen 2021). Thus, the divers managed to finance their operations without exhausting their own economies. However, as the group grew, so did their costs, necessitating the search for new avenues of finance.

Eventually, the group begun to reach out to the public, authorities, and organizations for financial help. They started to have signs along the shoreline that informed about the work being done and that asked for small donations from those passing by. They also contacted various companies whose waste was often found in the water to inquire if they would be willing to support the group in removing their waste. As with the authorities, the interest from these companies was low in the beginning. But with the growing public support for the group, the interest from companies also grew, as they saw the group as a possible good marketing opportunity. Finance started to become less of a problem following events in the summer of 2020, as will be discussed further in Chap. 7. Suffices to say here, the financial situation turned from complicated to one of solid finance, allowing the group to expand its activities, buy better gear, and even rent trucks equipped with cranes to haul large objects like cars out of the water (Fig. 6.7).

Fig. 6.7 Car being retrieved from the bottom along Söder Mälarstrand. (Photo: Cassiani Fotinos)

Covid-19: Volunteers Assemble

The relentless work of divers and rope pullers in Rena Mälaren gradually built momentum. Then the number of volunteers exploded shortly after the COVID-19 pandemic struck the world. With the pandemic restrictions that closed down large parts of many people's social lives, many became utterly restless at home. At the same time, Rena Mälaren's activities continued, as they were outdoors and it was possible to follow regulations and keep the work going. Thus, Rena Mälaren offered opportunities for social interaction and physical activity, while also accomplishing environmental work. This appealed to many people. Whereas earlier there may have been 10–12 people at a cleanup, there could now be 50 or more. The increase in participants was so abrupt and significant that a parallel between Covid-19 restrictions and the growing number of volunteers for Rena Mälaren could be drawn. What had started as two individuals in the summer of 2019 quickly turned into more than 100 after the pandemic, and by mid-2022 the number had grown to over 300.

The massive rise in volunteers was generally positive for the project, as it increased publicity and muscle-power to bring waste up. However, there were downsides, too. For example, safety became an acute concern. There is a significant difference between making sure that five or six rope pullers follow their diver closely compared to having 30 or more rope pullers on shore. To increase safety during cleanups with the large number of new volunteers, it was mandatory to participate in a pre-cleanup briefing where the leading diver and the dive leader went over what would be done on that occasion. They explained which diver would be paired with which rope puller, what area they would be diving in, when they would be in the water, and how the diver and the rope puller should communicate. Finally, everyone was given the phone number of the dive leader, should he/she not be within earshot when needed. At first, many divers and those organizing the dives were a bit nervous, as risks clearly increase with new, inexperienced volunteers. Still, they felt comfortable that all the participants listened carefully to instructions and did their best to follow them. In time, also the new recruits became experienced in the work. It was a stressful first couple of weeks. But as soon as most of the newcomers had taken part in a few sessions, things got back to normal, except that now the group was much larger, and the amount of waste they could pull up increased tremendously. Inger, a diver in her early 20s, summarized it this way:

Fig. 6.8 Rope pullers engaged in pulling waste from the water. (Photo: Joakim Hall)

> When all these new people first started to come, it felt like there were two groups—the inner circle and the newcomers. However, the enthusiasm these new volunteers have brought quickly breached this boundary, and now it truly feels like we are one group again, only bigger.

As mentioned earlier, Rena Mälaren was from the start meant to be an inclusive project, something that everyone should be able to feel comfortable joining. This standpoint became visible with the increase in the numbers from Covid-19 shut-ins. Not just single individuals turned up: whole families made it their weekend activity to join the divers in the work of pulling waste up from the water. As one father said, "The kids just love it. It is like the game at children's parties where you throw a stick at a piece of fabric and out comes a bag of candy. This is the same, but just much more exciting." (Fig. 6.8).

Organizational Structure

As the group grew it became evident that they had to formalize their organization and structure in order to gain more control over who could participate and how. This was also important to control the increasing inflow of money—sponsors wanted to know where their money went. Therefore, the previously informal group registered as a formal non-profit organization. The structure of the organization is democratic, with elected representatives and a leading group. Key decisions have to go through a vote during the annual meeting, or an extra meeting if the matter is urgent. Fredrik was the uncontested choice for a leader, and then another diver became the economic representative, and I was, as I mention in the method section, asked to take the role of secretary. Soon the managing group was extended to a number of five, with one person responsible for social media and another with an advisory function. Matters of concern for members are discussed during the annual meeting, and all members can bring up suggestions and concerns. Votes are decided with a majority principle. If enough members agree, suggested measures will be implemented (some changes need to be voted on during two meetings in a row if they concern changes in the organizational structure).

To be a formal organization has been beneficial for Rena Mälaren as it has allowed them to also apply for funding through more regular funding channels, other than donations. It has also given the group more credibility in the eyes of politicians and the public—as they now can be held accountable for their activities as an organization, not just as individuals.

More Waste Up, More!

The bottoms of Lake Mälaren that I have explored look like a large garbage bin with piles of waste everywhere. The two images below show a typical view along Söder Mälarstrand, one of the central waterways in Stockholm, at 3 meters below the surface. There are some larger objects such as car tires, scooters, and various cans. Beneath these objects is a mass of smaller, harder to identify bits and pieces. These are often parts from larger metal waste that has eroded and lost its structure, and thereby spread out along the bottom. One can see this by taking some in your hand: it will crumble into tiny particles, while there often still is an intact core of harder metal. These cores are often sharp and can easily make holes in diving gloves or suits. Another thing that becomes evident in these images is that the visibility is not great. In these images, it is less than one meter, and there are so many particles in the water that it is often a fog-like experience. It is dark, too, especially when the diver goes under a boat. If visibility was grayish before, everything immediately turns pitch black as you pass beneath a vessel (Figs. 6.9–6.10).

Figs. 6.9–6.10 Common scenes on the bottoms around Stockholm. (Photo: Oskar Kihlborg)

In the early days of the project, the trash scuba divers mainly searched for and removed car batteries. These are the most toxic items along the bottoms, as they contain a large percentage of the heavy metal lead.[6] Another important reason for not bringing more types of waste up was that there were not many divers initially, hence removing batteries took most of the divers' time. Batteries were also easy to return to waste management stations, which was not true of other types of waste such as car tires. However, with the increase in volunteers, divers, and rope pullers, as well as the growing desire within the group to expand and make a visible impact on the environmental-political situation in Stockholm and Sweden in general (and perhaps even further away), the group decided to start removing other types of waste as well. They kept a tally of how many kilograms of waste they removed from the bottoms around Stockholm: in early 2024, the total amount was above 348,000 kilograms and growing. Of this, 39,000 kilograms were car batteries (an inventory of typical objects found on the bottom may be found in Appendix A) (Figs. 6.11–6.12).

Thus, what at first was a search for a specific type of waste, such as car batteries, has now turned into a "remove all that you can find" operation. As Per put it,

[6] The heavy metal lead has proven to be extremely toxic to humans and other beings. For a study on what lead contamination of water does to humans, see: Daniel Renfrew's book *Life without Lead: Contamination, Crisis, and Hope in Uruguay* (2018).

Figs. 6.11–6.12 Car batteries collected from the bottom, overgrown with marine organisms. (Photo: Anonymous Diver)

> When we only searched for batteries, it was like mushroom picking: once you had spotted one, your eyes became adjusted to signs of batteries in the sediment. This might be a sharp corner sticking out of the sediment or a thin cable looping out of the sediment and then disappearing down into it with about 15 cm exposed—the handle some batteries have to make them easier to carry. Once you have spotted the pattern, a sort of urge comes over you to find more and more! It is almost as if you become angry and disappointed when you cannot find any more batteries, although, really, that is very good, as it means no batteries have been disposed of in this area. It is a very paradoxical feeling! Now when we remove batteries and other types of waste, it sometimes feels like a buffet, sort of an all you can muster to bring up kind of feeling. It is a thrill. It feels weird admitting it, as one should not want any waste in the water to begin with. However, removing all that waste is a very exciting feeling.

This excitement fuels the energy of the group, but can sometimes turn into an obstacle to the work. For example, when there are several divers diving simultaneously, it is important that they spread out to give each one a space to dive in. Normally, when conducting recreational dives, you dive in a pair of two to increase safety. This is not the case when you conduct trash dives because if you dive close to one another, you effectively remove visibility from each other, making the work much more difficult, and even more dangerous. The rope between diver and rope puller on the surface is the safety substitute that replaces a dive-buddy during trash diving. Furthermore, when planning the dive, the current is a critical factor to take into consideration as that will move the sediment particles along with it. Thus, if the first diver enters the water at the beginning of the dive site and the current is moving toward the site, then that first diver will destroy the visibility for the rest of the divers. Often the idea is to dive against the current, so that the current will move the particles away from the dive site. However, once particles are in the water, they will most often stay there for some time before settling back down to the bottom. Hence, if one or a few divers dive just before the others, chances are good that they will have removed the visibility for those following them. This is where the eagerness to find the most waste can become problematic. Since there is a sort of competition among divers to see who finds the most waste at a site, on a number of occasions divers have come to the site before the scheduled time in order to get into the water first. By the time the others arrive, visibility is already gone. This has resulted in annoyed reactions from the other divers. On the one hand, if competition means that divers and rope pullers work harder, it is an effective motivation. However, when the thrill of getting the most waste up becomes more important than the collaborative effort, it becomes a hindrance and source of conflict.

There is also the issue of what waste should be removed and what should be left in the water, out of concern for marine organisms. Some objects that have been in the water for a long time can turn into habitats for living marine organisms. If removed, this home is destroyed, which could mean death for some organisms. Kim de Wolff (2017: 24), writing about plastics in water, explains with reference to Mary Douglas (1966) that some waste in the ocean "is not simply human-made 'matter out of place' [but] it is also a matter of life (and death) as bodies come to live, travel, and die with plastic, and with plastic inside them." The same goes for scrap metal in the water. Many snails and clams attach to and live on these pieces. And since these metals are not toxic, should they really be removed? The question is complicated because, on the one hand, this waste has become a habitat for marine organisms, and therefore they provide a positive contribution to the marine environment (a similar situation is brought up in Chap. 4 when discussing the plastisphere). On the other hand, recycling this metal scrap rather than producing new metal is very energy efficient and saves a lot of resources elsewhere. For example,

Fig. 6.13 Sea stars found in the wheels of a large metal container brought up from the water in Gothenburg, Spring 2022. (Photo: Joakim Hall)

according to information from the authorities, the divers work on the assumption that producing one ton of new metal releases 2000 tons of carbon dioxide into the atmosphere. Instead, recycling one ton of old metal in Sweden releases only 400 tons of carbon dioxide. A "regular" car annually releases 3000 tons of carbon dioxide in Sweden. Thus, every 1.9 tons of metal the divers bring up from the water equates to the amount of carbon dioxide that a car releases each year in saved carbon dioxide emissions. This has led to the divers deciding to remove metal scrap unless it is overgrown with marine life. In the event some snails or clams are brought to the surface; the divers try to pick them off the metal and return them to the water with the hope they will find a new home (Fig. 6.13).

When life is found on the trash brought up, there can be surprises. Sometimes, so-called invasive species are found. According to the marine biologists in the group, invasive species are not just any foreign species, but species that have a negative impact on other "native" species in the local marine ecosystem. Following Swedish guidelines, these invasive species are not to be returned to the water, and they should be reported into the national art databank called "Artportalen." Rena Mälaren has its own page in this databank, and it is here they report if they find any living being defined as an invasive species. In the report they are supposed to define the species, note how many or much of it they have observed, and the location of their finding. This far Rena Mälaren has come across five species defined as invasive in Sweden: *Chinese Mitten Crab*, *Harris Mud Crab*, *Small American Waterweed*, *Dark False Mussel*, and *Zebra Mussel*. Far from all the divers and rope pullers have knowledge about these species, but there is a guide produced by one of the members as how to identify and then act on these species. The idea is that the work to monitor species in the marine world might become a bigger part of the diving project in the future when more divers learn about what species to care for, and what species to remove as a way of caring for the "native" marine environment.[7]

Moreover, most "invasive species" are introduced by humans, knowingly or by mistake. They can hitch a hike in the ballast water on ships, or simply hatch on to the bottom of boats. To avoid contributing to moving such species around, the divers make sure to clean their gear carefully after each dive. They should rinse their suit clean from any vegetation, empty water from the BCD on the spot, and preferably let the suit dry for five or more days between dives in Lake Mälaren and saltsjön (the Baltic Sea). These actions will minimize the risk of some species moving between Lake Mälaren and the Baltic Sea using the divers as a means for transport.

Toxic Water and Bottoms?

When collecting waste, divers encounter different types of bottoms. For example, bottoms at old boat marinas are often extremely muddy, with thick layers of sediment. A diver's arm can press right down into it without even feeling a hard bottom. When the sediment is disturbed it will immediately move up and mix with the overlying body of water, effectively

[7] There is a large scholarly discussion on "invasive species" (e.g., Khalil 2019; Orion 2015; Head et al. 2015; Wanderer 2020). This is, however, beyond the scope of the story I am telling in this book. But I want to acknowledge it here, and state that I do not find the definition between native and invasive species to be an easy on. Nor the terms on their own. This is, however, a discussion for another time and place. What I will note, however, is that deciding what life is invasive or native is a politics of deciding what life is allowed to live, and where. This is part of the aquabiopolitical relationship I argue exist between humans and the marine world. Even though the intention behind removing so-called invasive species is to safeguard the existence and life of so-called native species, it is still humans deciding about life or death in the marine world.

removing all visibility. Another type of bottom is found along Norr Mälarstrand and Söder Mälarstrand. These bottoms are sandy, without a thick layer of sediment on top. Here it is easier to remove the waste from the bottom as the sand remains on the seabed, or returns quickly, which means that the visibility is often better. In addition, sand does not function as a lid that holds things in place below it as sediment does. There are a lot of old chemicals and heavy metals concealed well below the sediments around Stockholm, and these can leak into the water if the sediment is disturbed. The sediment is thus a sort of archive of past chemical pollution (Radomska and Åsberg 2021: 7). Therefore, divers need to be careful when removing waste from areas with significant sediment, so that they only take the waste visible from above and not dig down.

There is a balance between not disturbing old pollution and how much of newer waste one should leave in the water, adding to the already existing pollution. For example, at the beginning of the diving project, some officials advised divers that car batteries can safely be left in the water, saying that they do not leak toxicants but remain intact for a very long time. However, that is not entirely true. Fredrik had several batteries tested to determine the pH level in the liquid present within the tubes containing the sulfuric acid which has a very low pH. The tests showed pH7, the normal pH for water. This means that the acid had leaked out, and that which was now in the batteries was nothing other than seawater. The contradiction between the official's statement and the results of Fredrik's test refutes the claim of the officials that human-made containers for chemicals and heavy metals are robust and will endure through time. Furthermore, they have arguably viewed water as a diluter, in that water effectively can swallow and make unpleasant substances "go away." Rena Mälaren has shown that both ideas are questionable.

Knowledge, Memory, and Responsibility as Aquabiopolitical Relations

I view waste in the water around Stockholm as a "public secret," along with Michael Taussig's (1999: 50) definition "as that which is generally known but cannot be spoken." I argue this in relation to the 1964 film *Sjövett* that I describe in the background section of Chap. 1, where the government instructed the public in Sweden on how to dispose of waste in the water by putting it in boxes full of stones or by making holes in aluminum cans so they would sink. From this film, Stockholm has become known as an environmentally conscious city, an image the local authorities are keen to promote. For example, in the Environment Program for 2020–2023 you can read:

> The City of Stockholm has been a leading environmental city for decades. The first environment programme was produced in 1976 and in 2010 Stockholm was chosen as the first European Green Capital. With the Environment Programme 2020–2023, the City of Stockholm is consolidating and developing its role as an international role model in global environmental and climate activities. (2020: 4)[8]

Waste in the water is an issue that is not in alignment with this green agenda. It has, therefore, been consciously forgotten about, and hence has become a public secret, meaning that it is well-known that waste has been dumped into the water. During my fieldwork, I have met many individuals who told me stories about how they once threw waste in the water, but would never do so now. They often only told these stories after hesitating. Their first reaction to the piles of waste brought up was astonishment: "Who can be stupid enough to throw all this into the water?" As it turns out, this could have been almost any earlier resident of inner-city Stockholm when they knew no better.[9] Similarly, Marietta Radomska and Cecilia Åsberg (2021: 2) write about the ocean and seas as "culturally marked as the spaces of 'forgetting': out of sight, out of mind." Because water is a forgotten place, or "no one's water," as Fredrik once put it, what lies on the bottom has traditionally been allowed to stay there, slowly leaking toxicants and polluting, and acting as a form of "slow violence" (Nixon 2011). This is an aquabiopolitical relation that has been allowed to continue thanks to the "forgetfulness" of the inhabitants of Stockholm. It is a "bad" relationship, as Liboiron (2021) puts it, a relationship that not only allows pollution to continue, but in a way normalizes such practices. If levels of pollution are unknown, or kept within certain limits, no one needs to be held responsible, and business as usual can go on, despite the consequences for life below water.

[8] https://international.stockholm.se/globalassets/rapporter/environment-programme-2020-2023_ta.pdf (Accessed 27 August 2023).

[9] By referencing the inhabitants of Stockholm, I hope not to make this "us" too general. I agree with Max Liboiron (2021) on the need to be precise about whom we speak about and to.

Threshold of Pollution

Alex M. Nading (2020) has stated the obvious in writing that it is unavoidable to recognize that today we live in a world enmeshed in toxicants and pollutants.[10] Even the water coming from the tap in Stockholm that residents drink is polluted to some degree. Authorities conduct tests on levels of pollutants in the water to ensure that toxic levels are below a threshold that supposedly could cause harm to human bodies. However, there are uncountable numbers of chemicals in the water, and many chemicals, or combinations of chemicals, do not have calculated thresholds at which they are harmful for humans. Liboiron (2021: 40) has written about a "threshold theory of pollution," which is paired with "assimilate capacity." This refers to an allowable threshold of pollution that nature can assimilate (dilute) by itself. According to Liboiron, this has not only permitted authorities and industry to pollute to a certain degree, but also given them the right to do so as long they keep within stipulated limits. We will return to this in Chap. 7. As Liboiron notes, authorities use the water as a sink and storage place for waste. This waste is now below the sediment in the form of chemicals and heavy metals, or it lies along the bottoms as car batteries, metal waste, and other items. Liboiron (2021: 57) further argues, "The threshold … is where policy, accountability, and responsibility come together." This is also where the divers push the authorities in Stockholm, asking how much and what waste should be allowed in the water. The government is well-aware of the risks of toxicants in the water. One especially telling example stands out:

A central waterway in Stockholm is known to have excessively high levels of chemicals and heavy metals below the sediments, namely, Karlbergskanalen. The high levels are the result of many industries in the area that emitted toxic elements into the water; these elements now rest below the sediment. Rena Mälaren had long wanted to dive here to see

[10] Another way of thinking about these processes is proposed by Michele Murphy (2018: 11) by using the concept "alterlife." She explains it as "the condition of being already co-constituted by material entanglements with water, chemicals, soil, atmospheres, microbes, and built environments, and also the condition of being open to ongoing becoming. Hence, alterlife is already recompiled, pained, and damaged, but has potentiality nonetheless. If life holds together tensions between violence and possibility, braiding the organic and inorganic, body and land, and resides in the indistinctions between infrastructures and ecologies, recognizing alterlife attends also to openness, to a potential for recomposition that exceeds the ongoing aftermaths."

what it looks like and, if possible, remove some of the waste from the bottom so as not to worsen the situation in the future. It took a long time to get permission, and a permit was only given to one diver. That person had to have a chemical-proof dive suit and a closed-circuit system for breathing. This was to ensure that the diver's body would not come in contact with the water. The diver was allowed to film and remove some waste, but only if it in no way disturbed the sediment. After the dive, everyone was pleased that the authorities had allowed them to see for themselves what was below, and remove some waste from the area. However, they were also greatly concerned. Inger said:

> If the water is so toxic that we must have special dive gear to be in it, then it does not feel very comforting to know that this water is part of the body of water from which drinking water is drawn for the inhabitants of Stockholm. The water is not taken from this spot, but water moves, and toxicants move with it.

Andrea Ballestero (2019: 409–415) has declared water a "vessel of toxicity" and "a human-made context." This is also accentuated by the activities of the divers. In a way, humans in Stockholm become part of the water they drink, including the water of the Karlbergskanalen. Furthermore, even if levels of toxicants are measured, less is known about how these toxicants impact humans when mixed together in what has been called a "cocktail effect" (Liboiron 2021: 96). The relationship between one chemical and a human body can be altered when other chemicals are added, making it difficult to predict how bodies will react to these chemicals when ingested. Here the divers' emphasis is on water quality for human consumption. But what their work makes visible, on a larger scale, is that waste in water makes the underwater world toxic and possibly hazardous for all beings, human and non-human, who live in it or from it.

If we take car batteries as an example. When these batteries crack open, they leach lead and acids. Lead is known to harm most organs and functions in a body (human and not), with effects on, for example, the kidneys, the cardiovascular system, or even reproductive capacity. Since lead has no biological function for animals or plants it bioaccumulate and thereby it affects the entire food web—from small organisms, to fish, and eventually, humans (Botté et al. 2022; Flora et al. 2006; World Health Organization 2021). Rena Mälaren estimates that approximately 15% of the batteries they collect already have cracked open, and emphasize the importance of retrieving the other 85% before they do too. Lead already in the marine environment will not disappear on its own, as per Article 12.2 in the Association of European Automotive and Industrial Battery Manufacturers (EUROBAT) 2006 guidelines for battery use, lead contamination in water needs chemical cleansing.[11] Lead will not just go away on its own but instead enter living bodies or get buried in the sediment and stored there for the future. Recent studies show levels of lead, mercury, and PAHS in fish in Lake Mälaren. PAHS in high level (as mentioned in the Introduction), mercury close to dangerous levels for humans, and lead in lesser concentrations (Swedén and Olsen 2021). However, that concentrations are low enough for supposedly safe human consumption does not mean they are low enough for the well-being of the fish and other marine life. As an example, in 2015 residents of Stockholm could read about an event where five swans had died from lead poisoning in central Stockholm.[12] The obduction indicated that lead was the reason behind their death. The source of the lead was not determined, but it could have been from them eating fishing sinks that had been stuck on the bottom. This did, however, draw attention to the damage lead can cause in the marine world.

A political view of water as a dump for waste (Liboiron 2021), as a place where waste and toxicants will go away (Radomska and Åsberg 2021), illustrates another set of aquabiopolitical relations. Water, seen as a resource, has been appropriated for human use in the form of a sink for unwanted things (Reno 2015). A threshold of how much waste can be tolerated in the water has been put in place when declaring that the levels of specific toxicants are below harmful levels for humans. But what about the marine organisms? When divers point out the risk to human health, the public reacts, and asks the authorities to take action. The idea of a threshold is important as it is an effective governing tool: if levels are below the threshold, then everything is supposedly fine. But are the pollutants within safe levels? Or is this "safety threshold" just something that is allowed in order to keep certain industrial and economic practices going, despite long-term environmental costs piling up along the bottoms of oceans, seas, and lakes? These are questions that the divers seek to make authorities respond to, and that we will consider further in the following chapters.

[11] https://www.comstedt.se/wp-content/uploads/2016/02/SDB-Sakerhetsdatablad_blybatterier.pdf

[12] https://www.svt.se/nyheter/lokalt/stockholm/svanar-dog-av-blyforgiftning

References

Bakker, K.J. 2010. *Privatizing Water: Governance Failure and the World's Urban Water Crisis*. Ithaca, NY: Cornell University Press.

Ballestero, A. 2019. The Anthropology of Water. *Annual Review of Anthropology* 48: 405–421.

Botté, A., C. Seguin, J. Nahrgang, M. Zaidi, J. Guery, and V. Leignel. 2022. Lead in the Marine Environment: Concentrations and Effects on Invertebrates. *Ecotoxicology 31* (2): 194–207.

Cornea, N., A. Zimmer, and R. Vérnon. 2016. Ponds, Power, and Institutions: The Everyday Governance of Accessing Urban Water Bodies in a Small Bengali City. *International Journal of Urban and Regional Research* 40 (2): 395–409.

de Wolff, K. 2017. Plastic Naturecultures: Multispecies Ethnography and the Dangers of Separating Living from Nonliving Bodies. *Body & Society* 23 (3): 23–47.

Douglas, M. 1966. *Purity and Danger: An Analysis of Concepts of Pollution and Taboo*. London: Routledge.

Drew, G. 2020. Political Ecologies of Water Capture in an Indian 'Smart City'. *Ethnos. Journal of Anthropology* 85 (3): 435–453.

Ecott, T. 2001. *Neutral Buoyancy*. St Ives: Penguin.

Flora, S.J.S., G. Flora, and G. Saxena. 2006. Environmental Occurrence, Health Effects and Management of Lead Poisoning. In *Lead*, 158–228. Elsevier.

Haraway, D. 1991. Science, Technology, and Socialist-Feminism in the Late Twentieth Century. In *Simians, Cyborgs, and Women: The Reinvention of Nature*. New York: Routledge.

Hayward, E. 2010. Fingeryeyes: Impressions of Cup Corals. *Cultural Anthropology* 25 (4): 577–599.

Head, L., B. Larson, R. Hobs, J. Atchinson, N. Gill, C. Kull, and H. Rangan. 2015. Living with Invasive Plants in the Anthropocene: The Importance of Understanding Practice and Experience. *Conversation & Society* 13 (3): 311–318.

Helmreich, S. 2007. An Anthropologist Underwater: Immersive Soundscapes, Submarine Cyborgs, and Transductive Ethnography. *American Ethnologist* 34 (4): 621–641.

Jue, M. 2020. *Wild Blue Media. Thinking Through Seawater*. Durham: Duke University Press.

Khalil, A. 2019. *Don't Shoot the Messenger! Invasive Species and Halting Biodiversity Loss*. https://abeautifulresistance.org/site/2019/10/22/dont-shoot-the-messenger-invasive-species-and-halting-biodiversity-loss. Accessed 10 March 2024.

Liboiron, M. 2021. *Pollution Is Colonialism*. Durham and London: Duke University Press.
Merchant, S. 2011. The Body and the Senses: Visual Methods, Videography and the Submarine Sensorium. *Body and Society* 17 (1): 53–72.
Murphy, M. 2018. Against Population, Towards Alterlife. In *Making Kin Not Population*, ed. A. Clarke and D. Haraway. Chicago, IL: Prickly Paradigm Press.
Nading, A.M. 2020. Living in a Toxic World. *Annual Review of Anthropology* 49: 209–224.
Nixon, R. 2011. *Slow Violence and the Environmentalism of the Poor*. Cambridge: Harvard University Press.
O'Hare, P. 2019. Waste. In *The Cambridge Encyclopaedia of Anthropology*, ed. F. Stein, S. Lazar, M. Candea, H. Diemberger, J. Robbins, A. Sanchez, and R. Stasch. https://doi.org/10.29164/19waste.
Orion, T. 2015. *Beyond the War on Invasive Species*. Chelsea Green Publishing.
Puig de la Bellacasa, M. 2017. *Matters of Care. Speculative Ethics in More Than Human Worlds*. Minneapolis, London: University of Minnesota Press.
Rademacher, A. 2015. Urban Political Ecology. *Annual Review of Anthropology* 44: 137–152.
Radomska, M., and C. Åsberg. 2021. Fathoming Postnatural Oceans: Towards a Low-Trophic Theory in the Practices of Feminist Posthumanities. *Environment and Planning E: Nature and Space* 5 (3): 1428–1445. https://doi.org/10.1177/25148486211028542.
Raycraft, J. 2020. Seeing from Below: Scuba Diving and the Regressive Cyborg. *Anthropology and Humanism* 45 (2): 301–321.
Renfrew, D. 2018. *Life Without Lead: Contamination, Crisis and Hope in Uruguay*. Berkeley: University of California Press.
Reno, J. 2015. Waste and Waste Management. *Annual Review of Anthropology* 44: 557–572.
———. 2016. *Waste Away. Working and Living with a North American Landfill*. Oakland, CA: University of California Press.
Rodineliussen, R. 2021. Caring for Water: Underwater Waste, Trash Diving and Publicity in Stockholm. *Kritisk etnografi—Swedish Journal of Anthropology* 4 (2): 67–81.
Swedén, J., and T. Olsen. 2021. Miljögifter i fisk från Vänern, Vättern och Mälaren. Utökad miljöövervakning i de Stora sjöarna. *Mälarens Vattenvårdsförbund*. https://vattern.org/wp-content/uploads/Rapport152_Miljogifter_fisk_2021_Stora_Sjoar.pdf
Taussig, M. 1999. *Defacement. Public Secrecy and the Labor of the Negative*. Stanford: Stanford University Press.

The Environment programme 2020–2023 Serial no.: KS 2019/1040 | Release date: May 2020 | Publisher: City Executive Office Contact person: Ebba Agerman | Illustrator Environmental Goals: Tobias Flygar https://start.stockholm/globalassets/start/om-stockholms-stad/politik-och-demokrati/styrdokument/environmentprogramme-2020-2023_ta.pdf

Thompson, M. 2017 [1979]. *Rubbish Theory. The Creation and Destruction of Value—New Edition*. London: Pluto Press.

Wanderer, E. 2020. *The Life of a Pest. An Ethnography of Biological Invasion in Mexico*. Oakland: University of California Press.

World Health Organization. 2021. *Lead Poisoning*. https://www.who.int/newsroom/fact-sheets/detail/lead-poisoning-and-health

CHAPTER 7

Looking into the Underwater World: An Attempt to Make What Is Below the Surface Visible from Above

The previous chapter showed the *how* in the work of Rena Mälaren, providing descriptions of trash divers and rope pullers, explaining financial constraints, and emphasizing sensorial aspects of this endeavor. In this chapter, the objective is to show how Rena Mälaren seeks to change the land-based bias of the way in which humans think about water (Raycraft 2020; Jue 2020). They do this by making waste below water visible and tangible to humans above the surface. Its members think that it is of key importance to see, feel, and smell the underwater waste to gain an understanding of what the waste does to the marine environment. For this reason, the group has made a number of installations that bring the spectator into a sensorial relationship with these waste objects. These installations have then been broadcast on social and traditional media to reach a larger audience than those who had seen the installations live.

We have characterized waste in water as a form of "slow violence" (Nixon 2011), in that it slowly leaks toxicants that accumulate and pollute the underwater world for a long period of time. The abovementioned actions of Rena Mälaren can be viewed as attempts to "work time" by making a "moral punctuation" that emphasizes the impetus to act now, before things become even worse. A moral punctuation, according to Chloe Ahmann (2018: 143–144), is "an explicit marking of time that condenses protracted suffering and demands an ethical response,

R. Rodineliussen, *Underwater Worlds*, Sustainable Development Goals Series, https://doi.org/10.1007/978-3-031-63370-6_7

eschewing the delays of political caution and the painstaking work of ensuring scientific certainty." When the divers make the slow violence visible and emphasize the importance of acting now, they point to the aqua-biopolitical relationships that allow this situation to continue. They do so in a bid to change current practices, on the governmental and public levels, with the hope that it is not too late to learn to live with and take care of the underwater world in a different way.

The chapter starts with an example of how the divers made a statement for clean healthy water with an underwater demonstration against the local government. This was in response to Stockholm's decision to continue dumping polluted snow into the water, a practice that had been going on for about 20 years. The importance of social and regular media for the divers as a platform to reach out and educate, but also for social organization and mobilization, can be seen in how the divers made art out of the waste from the water. With their art installations they collaborated with other environmental organizations and initiatives, such as Fridays for Future. An important role is played by images: what they can depict and how they are received, especially in the context of thousands of similar images. The divers also work to reach the younger generation and educate them for the future.

Underwater Demonstration Against Ongoing Pollution of Drinking Water in Stockholm

It is winter and the streets of Stockholm are full of snow. Cars drive on it, people walk in it, and waste gets tossed into it. The snow is first pushed to the curb, to lie as a border around the streets, absorbing plastic particles from car tires and the exhaust from vehicles driving past. The snow that once was white turns grey and black. This snow is then collected on trucks and taken away by the authorities to be dumped into Lake Mälaren. This is, in fact, against the law; but, the city routinely applies for and is granted special permission to discharge the dirty snow into the lake, claiming that there is no viable alternative.

On Rena Mälaren's Instagram from 4 December 2021, a statement appeared regarding this practice of dumping dirty snow into Lake Mälaren.

> The snow contains a number of toxic elements. From the Traffic Office's own estimations, it is possible to read that the dumping of snow yearly con-

tributes high levels of heavy metals, above 300 tons of microplastics, and close to 250 tons of road salt directly into our drinking water.

To protest such dumping, Rena Mälaren organized an underwater strike on 3 December 2021 at one of the locations where the dumping typically occurs. The Instagram post mentioned above included several images from this demonstration, shown below. The protest entailed four divers sitting on the bottom with signs reading "Stop the snow dumping." Over the course of the demonstration the divers on the bottom were filmed and live-streamed onto a screen on the surface. A communication device allowed the divers to address the crowd from below the water through a set of loudspeakers. Thus, people passing by could not only look at the moving images of the divers in the water and read the signs, but also hear the divers speaking to them about ending the practice of dumping polluted snow into the water, followed by loud *phsss phsss* sounds from them breathing through their regulators.[1] (Figs. 7.1–7.4).

The divers were below water to emphasize that the waste ends up there, right into the drinking water of the approximately 2 million inhabitants of Stockholm. When Rena Mälaren asked the authorities about the reasons behind this dumping, they answered that they did not have better ways of disposing of the snow. Rena Mälaren asked why it would not be better to pile the snow onto a soccer field, let it melt, and then collect the waste hidden in the snow. The response was that many toxic elements would leak into the earth when the snow melts. The divers questioned this logic, asking whether it was really better to just dump it into a source of drinking water.

In a group messenger chat that I described in the method section of the Introduction, the divers discussed this response from the authorities. One of the divers shared a video of how the authorities in Oslo, Norway, had solved the problem of dirty snow. It did not make the divers any happier with the way the snow was handled in Stockholm since it became obvious that other solutions were possible (the video is linked in the footnote

[1] The way the divers of Rena Mälaren act, by placing themselves in the environment under stress, is similar to how the leaders of the Maldives acted on the 17th of October in 2009 when signing a protest against the leading powers of the world and their inaction towards climate change, which soon will put the Maldives below water (Helmreich 2011: 1).

Figs. 7.1–7.4 These four images depict the underwater demonstration staged by Rena Mälaren. Top left: A truck caught in the action of dumping dirty snow into Lake Mälaren. All images except the top right are screenshots from two videos made by Oscar Khilborg. One video was also shown on SVT News. The top right image is taken by Oscar Khilborg

below).[2] In Oslo the authorities had built a cleanup facility especially for processing dirty snow removed from roads during the winter. They did so in order to keep the pollutants hiding in the snow away from the environment. In this video, one of the plant's employees explains the procedure and shows examples of how much waste and pollutants are found within the snow (in Norwegian). This video illustrates another way of taking care of the dirty snow and the quantity of pollutants the authorities in Stockholm dump into Lake Mälaren.

[2] https://m.facebook.com/watch/?v=2204929426425385&paipv=0&eav=AfZjG4o-VlAzFKLTRgU5OFRCZAxhllbUxjox%2D%2Dy_m-Sq6x1aThCFn-agDUrwQRgF654&_rdr. Accessed 1 August 2023.

This underwater demonstration is one of the divers' attempt to make what is below the water visible to those on the surface, through different senses. They do this to protest the long-term and ongoing use of water as a dump site for waste (Reno 2015; Rodineliussen 2021). The divers point out how the Stockholm authorities prioritize an easy and cheap solution for discharging the polluted snow over the health of the marine environment and, in the long run, humans too, as humans consume this water.

In Chap. 6 we learned about the vision of Stockholm to become increasingly environmentally friendly, and that the city views itself as, and has been awarded the title of, a green capital. When the divers point out that the authorities have dumped polluted snow into the water for about 20 years and plan on continuing this in the future, they expose the ongoing longitudinal violence done by the authorities to the marine environment. By making a spectacle of this practice with their underwater demonstration, the divers make a moral punctuation (Ahmann 2018). As one seasoned middle-aged diver by the name Bo reflected,

> If Stockholm is supposed to be a green, or even a blue, capital, then this cannot continue. We are actively destroying the amazing gift of fresh water. This city awards a water prize annually, while we dump waste into our own water. This is simply wrong.

The underwater demonstration was accompanied by a short video posted on social media by Rena Mälaren where they show footage of snow being dumped into the lake. Fredrik is seen crawling around in the water, moving waste and fill a bottle with water that was later sent for analysis (Fig. 7.5).

The image above, a screenshot from the video, depicts Fredrik sitting on a piece of packed dirty snow, watching the truck dumping waste into the water in the background. In the video you can hear Fredrik saying, "It is crazy! There must be other ways of disposing of the snow." As Oslo demonstrates, there are other ways, but they are costly. The video ends by pleading, "Please put an end to this madness!" Despite the divers' efforts, the authorities went ahead and applied for permission to continue dumping dirty snow, which they were granted for another five years. The divers were not surprised that the authorities decided to continue, but they plan to continue working in the coming years to stop the dumping.

The members of Rena Mälaren are aware that during the present environmental crisis, with threats of global warming and extreme weather, the

Fig. 7.5 Screenshot from the video where Rena Mälaren discusses dumping snow into Lake Mälaren

news about ongoing practices of water pollution (such as dumping dirty snow) can be overwhelming, resulting in people feeling incapable of making a difference.[3] However, Rena Mälaren works to publicize the problem and lead by example, demonstrating how individuals *can* make a difference. This is probably one of the key reasons for the great response the group has received. While they paint a grim picture of the past and the future, they simultaneously show what we, as individuals, can do to work toward another, more optimistic, version of that future. The past is invoked to start us thinking about the future (Ialenti 2020), turning the pressure of the moral punctuation into something positive: let us have fun and clean together for a healthier future. This education, mobilization, and outreach is mainly done via social and traditional media. Let us therefore look at how the group makes use of these media venues in their attempt to make what is below the surface visible to those on land—sometimes in ingenious ways.

[3] The word "crisis" is somewhat contested. I use it because it is used by the individuals with whom I have worked. However, as Latour (2017) asks, is the world really in an environmental crisis? The word "crisis" indicates, according to Latour, that there is an end to the current state of affairs, a point at which the crisis will end. But will this environmental crisis really end? Or have we reached a new normal?

Media as a Venue for Outreach

When Rena Mälaren began, they did not use any social media. Any publicity they received came through traditional local media outlets. By "social media" I mean platforms that individuals use to share content or communicate among themselves, such as Facebook and Instagram. Whereas "traditional media" refers to printed newspapers, radio, and TV. However, relatively early, members of the group decided to see what kind of impact social media would have on the group's expansion, and how they could reach more people with their message about the damaging relations between humans, waste, and water. For this reason, they created a Facebook page and an Instagram account that were linked, so that they only needed to publish on one of the platforms for the message to appear on both. The idea was to gain visible momentum by showing how they were working to clean the waters around Stockholm. The hope was that this would attract more individuals who would be interested in contributing to the project and thus build a group of followers online. As Inger said, "The more people we have behind us wanting the same thing, the easier it will be to convince politicians to act." (Figs. 7.6–7.7).

Figs. 7.6–7.7 A diver and two rope pullers in action on a snowy winter day in Stockholm in 2021. (Photo: Cassiani Fotinos)

At first, the group's images and videos gained a few "likes," but as the regular posts continued week after week a following started to grow, especially when photographs showed work being done despite the weather, as in the images above. With the construction of the City Hall replica (described below) likes for every post and update ran from the hundreds into the thousands. Alice Mattoni (2013) has related how in many countries activists who cannot access traditional media as a platform to speak from, have taken to social media to communicate their ideas.[4] In Sweden this is somewhat different. The trash scuba divers have indeed used social media to tell their story, however, traditional media has also shown an interest in covering the work of the group. Beginning with local media coverage, but soon spread to national newspapers, TV, and radio channels. Thus, even though social media played, and still plays, a significant role in creating the public image of Rena Mälaren, traditional media also played and plays a big role. Today the media presence of Rena Mälaren is what I would define as hybrid, consisting of social and traditional media as well as offline activity (Lokot 2018). This way of acting, both online and offline, and with social and traditional media, is similar to the operations of many social movements and organizations, such as the Occupy Wall Street movement (Juris 2012; Razsa and Kurnik 2012), activists in Syria (Rodineliussen 2019c), or the #MeToo movement (Uimonen 2020a).

The divers brought waste ashore, showing it offline in the form of sculptures, and online as images, displaying what the public would otherwise never have seen. They set up huge piles of waste along the waterside to present what they found along the bottoms, while sharing images of the piles on social media. This was effective in creating reactions of disgust at how much waste was in the water. However, as many people passing by commented, these piles were so discouraging and outright frightening that it was hard to relate to them. When members of Rena Mälaren realized this, they began to think of other ways to display the waste so that those walking by could embrace the problem, understand it, and think about how to change the situation, rather than being gripped by a state of revulsion. Long discussions ensued within the group to think of other ways of displaying the waste, but doing it in a less offensive and more engaging and action-oriented way. After some time, the idea of building

[4] This has been described in numerous accounts such as, for example, the Arabic Spring (Stepanova 2011; Allagui and Kuebler 2011), and the so-called Syrian revolution (Harkin 2013; Mcclellan 2013; Fuchs 2014).

art installations out of the waste was brought up. But what would such installations look like? Further discussions led to the conclusion that making replicas of well-known symbols of Stockholm and Sweden might be a fruitful method of being critical while adding a touch of humor.[5]

How and What to Build

The brainstorming for what these installations might look like and how and where to build them, mainly took place in Fredrik's home. He lives on a boat docked in the central area of Stockholm. We were sitting in the main room, with views of water all around us, and with the boat slowly rocking to the motion of the waves below. Those attending were mainly the more senior members of the group who had been part of the work since the earliest days, plus some new creative minds. The table was full of papers and pens, and people were occupied with sketching different ideas and designs while others searched the web—maybe others had done similar things elsewhere that could be used for inspiration? The group did not find any good installations to mimic, but there were some ideas, such as an elephant made out of car tires, which led Fredrik to call out, "Hey! Let's play with one of Stockholm's best known symbols. Let's build a replica of the City Hall and the Three Crowns!" The group quickly rallied around this idea and decided to make the tower and the City Hall out of car tires.

The original idea was to build the whole City Hall of only car tires, but then the group realized that it would be extremely complicated to make it solid enough not to fall down during the exhibition. Just building the tower from only car tires would be complicated enough. The construction

[5] Historian of technology, Andrea Westermann (2022), provides interesting insights into the role of plastics waste as a form of art and esthetics in her chapter "Against the Aestheticization of Technofossils," wherein she explains how plastics defined Western pop culture in the 1960s and 1970s, even its sonic history: "The sonic history of plastic added important nuances, too: think of *vinyl* as in vinyl records." Furthermore, she dives into how salvaged plastics from the environment were turned into artworks from the 1970s onwards. The way the divers from Rena Mälaren acts with waste can well be described as following a similar pattern, as part of "art in the Anthropocene." Importantly, this art is meant to impact viewers (Favero 2020), attaching an emotion of disgust (Ahmed 2014) to these objects, emotions that can inspire change in how individuals act with waste. For as Westermann (2022) explains, such art objects are not meant to become fossils, but are created to make people react against plastics pollution, and in a way, prevent it from becoming what Michell Bastian and Thom van Dooren (2017) calls the new immortals of our time. The art is meant to move individuals into action for the environment, and against plastics pollution.

was to be mounted on a barge so that it could be placed on the water in front of the actual City Hall. The barge would move with the waves, and car tires brought up from the bottom are not very stable as they have become soft from all the particles that have dissolved into the water. Hence, something had to keep them together. The group jointly came up with the idea of setting a long metal rod within the stack of car tires and tying the tires together to that rod with wire to increase stability and still allow some flexibility to compensate for the movement of the barge on the water. When the design had been finalized, the group began planning how to collect the waste needed, which would amount to several tons. It was decided that the city would be asked to store the waste during the spring as the group collected it, but that request was declined. Then the divers came up with the idea of collecting the waste along the shoreline around town, but still below the water. From there the tires and other waste could quickly be pulled ashore when they were needed for the construction. This was not an optimal way of doing it, as members felt bad leaving the waste in the water for an extended length of time, but since the group had no way of storing this amount of waste on land this was the only solution. Those pre-collecting dives were advertised on social media to start building up interest in the construction among the general public and possibly attract some more volunteers to assist in the work. The City Hall replica would be the first installation, but many others would follow.

Waste Installations

Rena Mälaren has now built several waste sculptures and art pieces, such as a replica of the City Hall, a Christmas goat, a Christian cross, a mini-version of the Eiffel Tower, Ariel the mermaid, and SpongeBob. More are being planned. These sculptures mock or align with certain relations, traditions, or commitments and connect with different groups in society at large. For example, the Christian cross was a collaboration with the Swedish Church, which wanted to make their engagement with the environment clear and visible. Therefore, they asked Rena Mälaren to assist them with an art piece for the environment during the Christmas celebrations of 2021 (Fig. 7.8).

For that event, Rena Mälaren built a cross filled with car batteries that was erected outside the Nobel Museum on the central square of the Old Town of Stockholm, close by the Stockholm Cathedral. On another

Fig. 7.8 The Christian Cross, full of car batteries. (Photo: Anonymous Diver)

occasion, still in collaboration with the church, the divers placed cars retrieved from the water on a graveyard in Stockholm to symbolize the remains of once living vehicles. These cars were allowed to stay in the graveyard for two weeks during the Halloween celebrations in 2021. The exhibit played with the idea that these cars were the living dead below the surface, and that they now had come up to haunt the inhabitants of Stockholm. The cars are of strong symbolic value for Rena Mälaren, as they are the carriers of the lead batteries that initially launched the group's work to clean the bottoms around Stockholm. The cars had yellow tape around them that marked the area. On the tape it was written *Brottsplats Mälaren* (Crime Zone Mälaren) as a pun directed at the now criminal act of dumping waste in water (Fig. 7.9).

The Eiffel Tower, on the other hand, was a collaboration between Rena Mälaren and the Fridays for Future movement. This sculpture acknowledged the failure to fulfill the commitments of the Paris Agreement, using a well-known symbol of Paris. The sculpture was created on 15 June 2021, five years after the agreement. Not a single country of the 196 that ratified the treaty has yet fulfilled the obligations to which they agreed. Here Fridays for Future and Rena Mälaren joined forces in critiquing the leaders

Fig. 7.9 Left: One of the cars parked at the graveyard; Right: The Eiffel Tower and Fridays for Future. (Photo: Joakim Hall)

of the world and their lack of action for the environment. The tower was built entirely from cigarette butts collected around town. Cigarette butts are found in large numbers on land and in water and emit many toxic elements into the surrounding environment. With the Eiffel Tower, time was again emphasized, namely, that politicians actively wasted time they no longer have. The slow violence enacted on water and the environment at large by politicians was exemplified by the symbol of Paris being built out of human toxic waste collected from the marine environment. In the Paris Agreement many promises were made and still not much has happened. Rena Mälaren and Fridays for Future are jointly trying to stress the immediacy of the situation, that action is needed now, not tomorrow or the day after.

The single-largest sculptures Rena Mälaren has built to date are the City Hall replica and the Christmas goat. The cross, cars, and the tower were all created in collaboration with other environmentally aware institutions, organizations, and individuals.[6] This collaboration presented a united front against the increasing pollution of the water and showed the importance of acknowledging that the environment needs acute attention now. If the cross and the tower were meant to create relationships with other actors engaged in environmental work, then the City Hall replica and the Christmas goat were made to point out gaps in the environmental policy and work of the Swedish government, with particular focus on Stockholm. The goat exemplified the story about the famous Gävlebocken, which is built yearly during Christmas in the northern Swedish town of Gävle.[7] The goat is traditionally set on fire and burnt down, no matter how much security is placed around it. Rena Mälaren decided it would be fun and effective to make a copy of the goat entirely out of trash, as a way to gain more followers and raise awareness. Speculations about whether this goat also would be surreptitiously burnt down created even more curiosity. It also led to the goat only being allowed to stand for two days and with around-the-clock surveillance due to safety concerns from the police.

The 7-meter-tall goat was constructed of no less than 700 car tires collected from the bottoms around Stockholm. The tires were attached to a wooden skeleton that provided stability and helped to form the correct shape. The goat was built and placed at a central square in the northern part of inner-city Stockholm, and it was possible to sense the construction

[6] By "environmentally aware" I mean individuals who believe it is important to change the way humans, individually and as societies, act with regard to the environment on planet Earth: people who engage in cleaning their surrounding from waste, or advocate for less use of fossil fuels, or the burning of coal. I use "environmentally aware" as an umbrella term to include a variety of actors and perspectives with one thing in common: they seek ways to battle climate change, environmental pollution, and the overuse of the earth's resources. For literature on other such approaches; Knox (2020), Dewan (2020), Loloum et al. (2021). However, Rena Mälaren's members do not always join because they initially want to work for the environment. Their reasons may be that they think it is "cool" to be associated with the group and perhaps be seen in media. However, as time passes, these members grow more and more "environmentally aware" by seeing the state of the bottoms around Stockholm and through the influence of the members with whom they spend time.

[7] The Gävle Goat is a giant version of the traditional Swedish Yule goat figure made of straw. It is built in the weeks leading up to Christmas and erected at the central square in the town of Gävle. The goat construction was built for the first time in 1966. The goat is usually attacked and burnt down. As of December 2021, it has happened 38 times. Although setting the goat on fire is an offense that carries a sentence of up to 3 months in jail, and no matter how great the risk of prison time or the number of guards put up around the goat to safeguard it, it usually goes up in flames.

before you could see it due to its strong odor. More than one resident living close by mentioned the smell, and seemed happy the construction would only be present for two days. The security concerns of the police were solved by having one of the divers sit in his car all night and guard the construction. The guarding diver was Phil, a man in his early 20s, who volunteered to stay through the night as he was curious to see how people would interact with the sculpture when they thought no one was watching. It was good that he stayed for, as he told me the morning after the first night:

> So many strange things happened during the night. I thought I would just be sitting in my car, listening to music and sipping my coffee. But no! First a group of young men, who obviously had been drinking, came by with music and we all danced around the goat. When they left things became quiet for some time, but then I noticed a shadow on the other side of the fence, and a man suddenly walked away with one of the tires. When I confronted him, he said, "What is the problem, it is only waste. I want one of them." I told him to go find his own waste, that if he wanted some of "ours" he should ask first. Then, in the morning, I got to talk to a group of pensioners on a morning walk who were amazed at the goat and the stories I told them of how these tires were collected and why.

People interacted with the goat in all kinds of ways. One person wanted a piece of waste merely because it had been part of the goat. It appears that people in Sweden have a special relationship with this Christmas symbol. Moreover, the trash replica was named the "Damn Goat" (Jävla Bocken), which is a play on words in Swedish. This is because Gävle and Jävla sound similar in their pronunciation, the first being a town and the other a curse—a fitting relationship between the straw original and the waste copy (Fig. 7.10).

The construction of the City Hall replica was the occasion when the trash scuba diving project really took off and started to be seen by the public and politicians as a significant force. It became known as a group that got things done and had the means to influence politicians and organizations on matters regarding waste in water. The City Hall structure took over two weeks to build, with the divers experimenting with the design along the way. It was not as easy to attach the tires together as the group had imagined on Fredrik's boat. Since all tires were of different sizes and states of decay, they needed to be puzzled together to make a

Fig. 7.10 Left: The "Damn goat"; Right: The City Hall replica. (Photo: Author)

somewhat coherent form and also give stability to the growing tower. The group had made good use of social media to advertise it prior to the construction. This had attracted the attention of traditional media that had asked to cover not only the exhibition but also the process of constructing the replica. Thus, journalists found their way to the construction site almost daily. They talked to the members of the group about their aims with the construction and about the project in general. Media also observed the construction work and the continuous diving that occurred around it as the divers brought up the stored waste lying in wait below the surface.

In the end the construction grew to eight meters tall and weighed more than 12 tons. It was floating on a barge and moored outside the actual Stockholm City Hall for two sunny summer days in 2020. A band played music among the trash on the barge. The music and the smell made people passing by stop and stand for a while, looking at the construction, smelling the odor, and enjoying the music. At this point it was possible for the divers to approach these individuals in a relaxed and unhurried way, to talk about why the installation was built, what the group hoped to achieve, and how it was possible to assist the project—as a diver, on land, or by other means—to help realize these goals. By attracting those walking past by what they could see, but also with the music and the smell (freshly collected waste has a pungent odor), people engaged with the waste sculpture through multiple senses. Thus, with sensorial stimuli other than visual cues, new ways were found for people to be affected by these environmental issues (Despret 2004: 131).

Scientists and Activists Joining Forces

In preparation for the City Hall exhibition, Rena Mälaren decided to establish closer ties with the scientific community and make use of their knowledge of waste in the water. I was delegated to contact the plastics scientists from Trondheim and ask if they would be willing to provide the divers with specific information about the impact of car tires on the marine environment. Rena Mälaren wanted this information about car tires because the sculpture would be built primarily of tires. I contacted Andy Booth and Lisbet Sørensen (from Chap. 4), who were happy to assist. In fact, they were currently writing a paper on the impact of car tires on the aquatic environment. Andy said,

> This is really timely! I am working on it at this moment. Please let me know how the information is received, and send me a few images of the exhibition. We can use that to show our institute that we are part of outreach work.

The information Andy and Lisbet supplied was printed on a large banner that was hung on the side of the sculpture (shown below). The banner read:

- Tires and tire wear particles (TWP) contain a wide range of organic and inorganic chemicals that can leak out into the environment over time.
- In particular, UV stabilizers such as benzothiazole are present in high concentrations. These provide tires with protection from sunlight that causes them to degrade.
- Chemicals and metals present in vehicle tires can leak out into water, depending on their solubility.
- This is dangerous because: benzothiazole and its derivative chemicals, as well as the metal zinc, which also is present in tires in very high concentrations, are both very water soluble and toxic to aquatic organisms.

- Tires contain a large number of other organic chemicals, some of which are known to be carcinogenic (e.g., polycyclic aromatic hydrocarbons (PAHs). The amounts of these chemicals present in tires sold in the EU are regulated to levels considered to be safe. However, the fate of these chemicals when tires and TWPs end up in the environment are unknown.

The banner was very helpful as members of Rena Mälaren explained to those walking by why car tires are so harmful when in the water. Pointing to this list, signed by two prominent scientists from a famous research institute provided legitimacy to the argument that car tires are toxic to the underwater world.[8] The points on the banner refer back to the discussion earlier regarding thresholds for pollution and the uncertainty of how chemicals act in specific contexts and in combination with other chemicals (Liboiron 2021). The collaboration between divers and scientists is an example of how media publicity can benefit a cause and attract others with similar interests. Although Andy and Lisbet would have helped the divers even without the outreach capacity of Rena Mälaren, it was a bonus for them to become visible and associated with the environmental work of the scuba diving group. As Andy said, "of course we want to help with this cool environmental activism art." (Fig. 7.11).

Creating New Relations

Before constructing the City Hall replica, Fredrik had tried for over a year to reach the Environmental Mayor of the City of Stockholm (*Stockholms stads miljöborgarråd*) to discuss the possibility of joining forces in the battle against waste in the water. He received no response. However, with the

[8] Rena Mälaren has continuously stated that car tires cause harm to the marine environment and that they can leak carcinogenic substances into the water. In mid-2022 an email came from one of the leading figures in the Swedish car tire industry, asking Rena Mälaren for references that proved that car tires are harmful for the environment. The man acknowledged that, of course, tires should not be in the environment, but he was not happy about how much emphasis Rena Mälaren put on this. He even provided a study of his own that showed that car tire particles used on soccer fields are not dangerous to soccer players (the study was sponsored by the company the man works for). This was met with many laughs among the divers. For, as they pointed out (and also replied to the man), no one ever claimed that the toxicity of car tires could jump through soccer shoes. But the toxicants within tires can leak out into the water, and through that medium, enter human or non-human bodies and cause harm. The man did not reply.

Fig. 7.11 Left: The banner with information about car tires impacting the aquatic environment; Right: A closer look at the pile of car tires making up the tower. "Atlantis 2" is named after the legendary underwater city Atlantis. (Photo: Author)

enormous media attention the construction began getting, things changed. The divers were interviewed on prime-time TV by one of the leading media outlets, TV4 news, and most local newspapers, some national, wrote about the event. It became a huge media circus. At this point the Environmental Mayor of the City of Stockholm called Fredrik directly and a meeting was arranged. The result was that the City began assisting Rena Mälaren by collecting the waste that the group brought out of the water. This ended having to spend hours after each dive moving the waste to the waste management stations: now the City did this. In this way, Rena Mälaren managed to create a relationship between government, waste, and water in Stockholm, while also working on a similar relationship with the larger public. From this we see that when aquabiopolitical relations are made visible, it is sometimes possible for the scuba divers to push for and achieve change in attitudes toward water.

Paolo Favero (2020: 142) writes about the communicative capacities of art, based on his research among artists and video makers in India. He states that "the meaning of art today is to be found not in the work of art itself, but in the relations, it generates." This argument is helpful in thinking about the role these sculptures and pieces of art played in the divers'

Fig. 7.12 Left: The mermaid Ariel; Right: SpongeBob. Both were created by the artist Herr Nilsson. (Photo: Anonymous member of Rena Mälaren)

communication with the public and the authorities. The sculptures provoked thoughts about what we as a society do to the environment. Thus, they create a collaborative nexus between trash scuba divers and the authorities, while emphasizing the relationship between humans, waste, and water (Fig. 7.12).

The sculptures functioned as a critique of governmental practices, depicting official symbols of Stockholm and Sweden, while the trash art exploited the bond between the public and well-known fictional figures, such as Ariel the mermaid and SpongeBob. These pieces look like their original, but with a twist that emphasizes the waste issue. Ariel is an example. When she first appeared in the 1989 film, she was shown using a human fork as a comb, not understanding what to do with it. Here, she is instead placed among toxic car batteries—also human-made and foreign to the underwater world. What will she do with them? Moreover, Ariel stands in front of the royal castle in central Stockholm, turning her back on the historical power of Sweden while raising her hands high and giving her middle finger to the world. Ariel, a being from the underwater world, is here bringing a message to the world of humans, and it is not a message of friendship and affection. Rather, it is a message saying that what once was a relationship of mutual care has turned to disrespect where humans harm the beings below water.

Such a relationship has been described by Veronica Strang (2021), who argues that "water beings" traditionally were worshiped for their other-than-human capacities, and that the positive relationship between humans and water beings assured sustainable ways of life. Later, as humans began worshiping more human-like deities, human-centered practices led to the situation that Ariel now protests against. Paula Uimonen (2020b: 202) has also analyzed water spirits, in her case, the "Lake Goddess" and her material manifestation in art and literature. Focusing on the African writer Flora Nwapa, Uimonen illustrates how writing about the Lake Goddess is "not a static recollection of the past but a social commentary on the present, with a view to shaping the future." This view collates with the idea behind the sculpture of Ariel, that is, that the past is full of wrongs, and that we should now try to correct them for the future.

The SpongeBob sculpture, on the other hand, embodies the properties of a sponge, the ability to absorb, as his body is made up of waste. The SpongeBob series originated with the 1989 unpublished educational book, *The Intertidal Zone* by Stephen Hillenburg, meant to teach students about the underwater world. The series became popular worldwide, and as of March 2022, a 14th season started production. These pieces of waste art solicit empathy from the public. By showing childhood stories enmeshed in waste, they reach out for help in stopping waste from entering the water to avoid hurting the beings of the underwater world. They seek to shift the perspective in thinking about water from above to below. If this shift can be made, the divers believe that humans might become more careful with how they treat water.

Images as Windows to the Underwater World?

When Rena Mälaren posts pictures of these installations and pieces of art on their social media platforms, they do so with the hope that *images* can assist in making people on the surface understand how waste impacts the underwater world. Images need to be understood within a particular context. A viewer looks at and relates to what is depicted based on a preconceived understanding of the subject that the image displays and the context the viewer comes from (Taylor 1998; Butler 2010). The meaning of an image is also not fixed. McDougall (2005) has said that the meaning of an image can be unruly and has concluded that much depends on its written or oral framing when arriving at its "meaning." Karen Strassler (2020: 13–14) writes that "conceptualizing *all* images as unfolding events enables us to see them as contingent and politically consequential processes in their own right." Furthermore, Favero (2017) states that images can no longer be taken as proof of the world captured in a frame because images now are mediated and seen as framed by the photographer, rather than as a piece of reality (Butler 2010; McDougall 2005).

I have elsewhere shown how images can sometimes become part of too large a context and by their sheer number, become overwhelming for the receiver, so that the receiver is unable to act upon what is being shown (Rodineliussen 2019c). This has been described as "compassion fatigue" (Moeller 1999), a situation in which viewers cannot transform emotions into action because they are overwhelmed by those emotions. This continues to be an issue with images depicting waste in water. The number of such images is enormous and they reach viewers in countless ways, making it increasingly problematic with regard to how viewers relate to images showing waste in the marine environment. For example, a quick google search for plastics in water gives you 1,210,000,000 hits in less than one second. This huge number of images creates a situation where new images easily can merge with those already present and thus have a minimal impact on viewers. Hence, the public needs more than images if they are to relate to what they see and to start caring. For this reason, the trash sculptures and art created by Rena Mälaren are of exceptional importance. Not only images, but live interaction with people, online and in person, gets their message across.

Raycraft (2020: 315) wonders about the ability of visual media to represent the underwater world experienced by scuba divers in a discussion with Merchant (2011), concluding, "Media cannot entirely capture the holistic sense of immersion that comes with actually *being* underwater." Merchant (2011) emphasizes the importance of bodily immersion in the environment if one is to understand it. I agree with them both when it comes to understanding what it means to be in and understand the underwater world as a scuba diver (Rodineliussen 2019a,b). Nevertheless, in the quest to understand the amount and impact of waste underwater, media such as photographs can do much to tell the story, especially when paired with exhibitions of what these images display. Allowing viewers not only to see images, but to observe the physical objects taken up from the water, engage with their smell, and even touch the by now slimy surface of these objects establishes a relationship between person, waste, and water. The hope is that this relationship can change the way individuals perceive and interact with waste and water.

While this may not bring about a total understanding, it might result in a compromise. A compromise is not a failure, but a negotiated reality. It is a starting point from which one can move forward (Liboiron 2021). The sculptures and images created by the members of Rena Mälaren can be seen as a starting point to rethink how "we," the inhabitants of Stockholm, relate to waste in water. These waste objects are in a way both below and above water, in their material capacity of being underwater waste taken ashore. With their physical installations the divers evoke reactions among onlookers that can be transferred to the images displayed online, allowing a moral punctuation to be sensed by the public.

Being Visible Is Being Influential

The outreach work done by Rena Mälaren had several tangible effects. For one, many individuals that passed by on cleanups would stop and tell the group, "Thank you for all your work to clean up the water. We have seen you in the news and we follow you on social media." On other occasions those passing by would jump in and lend a hand with the cleanup in progress regardless of whether they were wearing clothing appropriate for close contact with muddy tires and other waste. Such individuals would often respond that they could always wash their clothing, but they wanted to help out when they had the opportunity to do so. Several of these impulsive volunteers later became regular participants in cleanups.

This growing support by individuals just walking by, and the acknowledgment that the work was being seen and appreciated, all helped spur the group members forward. Another direct impact on the work was the increasing ease with which the group attracted financial support from both individuals and organizations. The group could now ask individuals to sponsor their project by donating small amounts to the group's bank account. Many of those who stopped did send in a contribution, especially on cleanups days, when the group had a sign with information and a request for funds on the sidewalk. These donations often came with a statement along the lines of: "Hi! I just want to say that I just sent some money. Thank you for all your important work!" However, the real economic change came when companies started becoming interested in collaborating with Rena Mälaren. Some found that the project's association added positive corporate visibility and environmental appeal to their brand and gave Rena Mälaren a donation in exchange for being listed as a sponsor on the project's website.

Since their inception, Rena Mälaren has tried to attract the interest of companies whose products are often found in the water, asking them to help fund the work of retrieving these products, and preventing them from polluting further. At first, such interest from companies was close to zero. However, with the increasing visibility and popularity of the diver group, interest from these companies slowly grew until they, too, started to help Rena Mälaren financially. It was not only the possibility of good publicity that made these companies come around, but often the risk of bad publicity that eventually had this effect. For example, prior to the City Hall installation the divers contacted a number of companies that rented out electric scooters in Stockholm. The companies were told that many scooters pulled from the water would appear in the exhibition, and that it might look better for these companies if they were part of the cleanup work, to which the companies agreed. The offer of good and the possibility of bad publicity convinced these otherwise reluctant companies to become sponsors of the diving group's work and to assist, at least financially, in the work of cleaning the bottoms around Stockholm.

The increase in visibility and growth in followers on social media also led to the dive group having to think carefully about which companies to work with. For just as images of waste in water can tell a particular story that affects viewers in specific ways (Favero 2017), so too does which companies the group is associated with. For example, several tobacco companies expressed interest in sponsoring Rena Mälaren, something the group declined. Although the money was attractive and would be invested in environmental work, the group did not want to be associated with tobacco companies and their overall impact on the environment and human health. Thus, even though funding is of critical importance to the group, they are careful to make sure they can support the sponsorships they accept and still remain independent and able to work in a way they deem fit.

Reaching the Young

We mentioned earlier that Rena Mälaren held obligatory pre-cleanup meetings prior to dives in response to the increase in volunteer numbers following Covid-19 restrictions and the group's growing visibility in various types of media. These meetings often began with Fredrik giving a short talk on the goals of Rena Mälaren, recounting some achievements, and announcing coming events. In these opening remarks he often talked about the younger generation and the importance of reaching them in the hope that they will be more receptive toward the group's message that we need to care for our planet in order for it to care for us. As Hannah Knox

(2020: 128) noted when writing about climate change, "We can't change the past, but we can change the future." Toward this end, the group was fortunate in being visited by one of the largest TV news show for children and young adults in Sweden, *Lilla Aktuellt.* The program translates recent news into a shorter format in a language that young people can better understand, to help this group to follow the news of Sweden and the world. *Lilla Aktuellt* also features brief reports on topics they deem of special interest to their audience. One such report was from a cleanup dive with Rena Mälaren in May 2020 where they primarily interviewed some of the children present.

At this dive, my own children, Lili and Leo, were present (as they were on many other cleanup occasions), and the reporter asked their opinion on the work of Rena Mälaren. Lili, eight years old, answered, "The work helps nature a lot. I wish everyone would do this type of work and stop throwing waste in nature." And Leo, four years old, said, "If they do not dive, then they will not help nature, and then nature might die." The reactions to these statements have been many. Several individuals I met on later cleanups mentioned that their children had seen the show and urged their parents to look for more information about Rena Mälaren and how to join in future cleanups. One father told me,

> It was invigorating to hear the story told from the perspective of young children. They are so straightforward! If adults say, "Nature is in danger of being harmed," then kids say, "It will die." This honesty is refreshing and exactly what we need.

Rena Mälaren seeks to lead by example—not only to argue for change but to bring about the future they imagine. In Sweden, it is a common saying among parents that children will not do what they are *told,* but what their parents *do.* Thinking along the lines of Lili's and Leo's statements, it is clear that they have learned from Rena Mälaren's model and have been inspired to take care of the environment around them. This is one of the main critiques of world leaders by the Fridays for Future movement—that those leaders only talk. Not surprisingly, many of the Fridays for Future members are also active in the work of Rena Mälaren, as the group "puts their money where their mouth is." They walk the talk.

Is It Possible to Change One's Perspective on Water from Above to Below the Surface?

Terence Wright (2008: 88), Horst and Miller (2012), Miller (2023), and others, have argued that visual media play an important role in the way the world is understood. Accordingly, Rena Mälaren has attempted to tell a story about a waste-ridden underwater world by using social and traditional media. In doing so, the group seeks to shift the perspective on water from one above to one from below. The way the group has engaged with music, smell, and imagination, combined with images and videos, work around the constraint of spectators being above water—a hybrid approach that engages a multitude of senses simultaneously (Lokot 2018). When one stands next to a massive pile of car tires full of mud, fresh from the water, the odor rises and envelopes you. The very peculiar smell is heavy and fills your nose, even your mouth. The smell and the taste lingers after you have walked away, clinging to you for minutes afterward, and reminding you of your encounter with a material representation of the waste in the underwater world.

Smell is often seen as a token of water quality (Scaramelli 2013), and as such, is an effective sensuous stimulus. However, the way these dredged-up tires smell is not simply "bad," they smell alien to our noses—like a creature rising out of the water. One might recall such a smell from one's childhood if you played with mud from the riverbank. Thus, this scent is not unfamiliar to most people. Yet, this odor is coming from familiar objects, so clearly not from the underwater world. The fact that it is human-made creates a confused feeling. This confusion is important, as it makes you think, relate, and wonder how these "land objects" can smell like "water objects." This does not seem right. Moreover, they are not just any type of "land objects"; these are toxic. This particular body of water is where drinking water for the inhabitants of Stockholm is drawn. Now a circle starts to close: waste moves from land to water, toxicants leak out, they then travel back to land in the water humans drink. Humans thus become together with the waste when the toxicants enter their bodies (Nading 2020). This understanding shifts the perspective on water from one above to below the surface. If not all humans can enter the water physically, they can do so mentally. The trash sculptures and images facilitate this awareness.

The story of waste operates over time, and so another mental dimension is opened—the one of memory. Many individuals I met during my fieldwork told stories from their childhood while looking at the trash sculptures. Such as this one:

> It is crazy to look at these enormous sculptures of waste from the water. And I do not want to imagine that all of this actually came up from the water just here. However, when thinking carefully, it is not that strange. I remember times when my father and I would take a small rowboat with an old washing machine or something similar, row out during night, and just tip it overboard. That was how it was done. Or when a car got old or broken, it could be parked on the ice during the winter, so that it would sink as soon as the ice melted in the spring.

But are these only practices of the past? Unfortunately, they are not. Rena Mälaren collects many contemporary objects, such as electric scooters that cannot be dismissed as old sins (Rodineliussen 2021). Because objects from the water are both old and new, one sees that these are longtime—and still ongoing—relations between humans, waste, and water. Such aquabiopolitical relations become visible in the moral punctuation made by Rena Mälaren. These relations become clear if one thinks of water, waste, and humans from an underwater perspective.

References

Ahmann, C. 2018. It's Exhausting To Create An Event Out of Nothing: Slow Violence and the Manipulation of Time. *Cultural Anthropology* 33 (1): 142–171.

Ahmed, S. 2014. *The Cultural Politics of Emotion*. New York, NY: Routledge.

Allagui, I., and J. Kuebler. 2011. The Arab Spring and the Role of ICT's. *International Journal of Communication* 5: 1435–1442.

Bastian, M., and T. van Dooren. 2017. Editorial Preface. The New Immortals: Immortality and Infinitude in the Anthropocene. *Environmental Philosophy* 14 (1): 1–9.

Butler, J. 2010. *Frames of War. When Is Life Grievable?* London: Verso.

Despret, V. 2004. The Body We Care for: Figures of Anthropo-zoo-genesis. *Body & Society* 10 (2–3): 111–134.

Dewan, C. 2020. Living with Toxic Development: Shipbreaking in the Industrializing Zone of Sitakunda, Bangladesh. *Anthropology Today* 36 (6): 9–12.

Favero, P. 2017. "The Transparent Photograph": Reflections on the Ontology of Photographs in a Changing Digital Landscape. *RAI, Anthropology & Photography* 7: 1–21.

———. 2020. *Image-Making-India*. London: Routledge.

Fuchs, C. 2014. *Social Media: A Critical Introduction*. London: Sage Publication.

Harkin, J. 2013. Is It Possible to Understand the Syrian Revolution through the Prism of Social Media? *Westminster Papers in Communication and Culture* 9 (2): 93–112. https://doi.org/10.16997/wpcc.168.

Helmreich, S. 2011. Nature/Culture/Seawater. *American Anthropologist* 113 (1): 132–144.

Horst, H., and D. Miller, eds. 2012. *Digital Anthropology*. London: Bloomsbury.

Ialenti, V. 2020. *Deep Time Reckoning. How Future Thinking Can Help Earth Now*. Cambridge, MA and London: The MIT Press.

Jue, M. 2020. *Wild Blue Media. Thinking Through Seawater*. Durham: Duke University Press.

Juris, J. 2012. Reflections on #Occupy Everywhere: Social Media, Public Space, and Emerging Logics of Aggregation. *American Ethnologist* 39 (2): 259–279.

Knox, H. 2020. *Thinking Like a Climate. Governing a City in Times of Environmental Change*. Durham and London: Duke University Press.

Latour, B. 2017. *Facing Gaia. Eight Lectures on the New Climatic Regime*. Trans. C. Porter. Cambridge: Polity.

Liboiron, M. 2021. *Pollution Is Colonialism*. Durham and London: Duke University Press.

Lokot, T. 2018. #IamNotAfraidToSayIt: Stories of Sexual Violence as Everyday Political Speech on Facebook. *Information, Communication & Society* 21 (6): 802–817.

Loloum, T., S. Abram, and N. Ortar. 2021. Politicizing Energy Anthropology: An Introduction. In *Ethnographies of Power: A Political Anthropology of Energy*, Berghahn (EASA Series). New York and Oxford: Berghahn.

Mattoni, A. 2013. Beyond Celebration: Toward a More Nuanced Assessment of Facebook's Role in Occupy Wall Street. Hot Spots. *Cultural Anthropology Website*. https://culanth.org/fieldsights/beyond-celebration-toward-a-more-nuanced-assessment-of-facebooks-rolein-occupy-wall-street. Accessed 31 October 2021.

McClellan, C. 2013. Presidents at Work: Shop Displays of Political Portraiture in Prewar Damascus. *Visual Anthropology Review* 29 (1): 16–29. https://doi.org/10.1111/var.12001.

McDougall, D. 2005. *The Corporeal Image*. Princeton: Princeton University Press.

Merchant, S. 2011. The Body and the Senses: Visual Methods, Videography and the Submarine Sensorium. *Body and Society* 17 (1): 53–72.

Miller, D. 2023. Digital anthropology. In *Open Encyclopedia of Anthropology*, ed. F. Stein, S. Lazar, M.Candea, H. Diemberger, J. Robbins, A. Sanchez and R. Stasch. https://doi.org/10.29164/18digital.
Moeller, S. 1999. *Compassion Fatigue*. New York: Routledge.
Nading, A.M. 2020. Living in a Toxic World. *Annual Review of Anthropology* 49: 209–224.
Nixon, R. 2011. *Slow Violence and the Environmentalism of the Poor*. Cambridge: Harvard University Press.
Raycraft, J. 2020. Seeing from Below: Scuba Diving and the Regressive Cyborg. *Anthropology and Humanism* 45 (2): 301–321.
Razsa, M., and A. Kurnik. 2012. The Occupy Movement in Žižek's Hometown: Direct Democracy and a Politics of Becoming. *American Ethnologist* 39 (2): 238–258.
Reno, J. 2015. Waste and Waste Management. *Annual Review of Anthropology* 44: 557–572.
Rodineliussen, R. 2019a [2017]. Visual Methods to Study the Underwater World: Scuba Divers and a Sensorial Experience of Water. *Anthrovision* 5(2): 1–14.
———. 2019b. Ethnographic Exploration of a Sensorial Underwater World. *Irish Journal of Anthropology* 22 (1): 184–189.
———. 2019c. Organizing the Syrian Revolution—Student Activism through Facebook. *Visual Studies* 34 (3): 239–251. https://doi.org/10.1080/1472586X.2019.1653790.
———. 2021. Caring for Water: Underwater Waste, Trash Diving and Publicity in Stockholm. *Kritisk etnografi—Swedish Journal of Anthropology* 4 (2): 67–81.
Scaramelli, C. 2013. Making Sense of Water Quality: Multispecies Encounters on the Mystic River. *Worldviews* 17: 150–160.
Stepanova, E. 2011. The Role of Information Communication Technologies in the Arab Spring. *Implications Beyond The Region Ponars*. Eurasia Policy Memo. Institute of World Economy and International Relations (IMEMO). *Russian Academy of Sciences* 159: 1–6.
Strang, V. 2021. Elemental Powers: Water Beings, Nature Worship, and Long-term Trajectories in Human-environmental Relations. *Kritisk etnografi—Swedish Journal of Anthropology* 4 (2): 15–34.
Strassler, K. 2020. *Demanding Images: Democracy, Mediation, and the Image-Event in Indonesia*. Durham, NC: Duke University Press.
Taylor, J. 1998. *Body Horror. Photojournalism, Catastrophe and War*. Oxford Road, Manchester: Manchester University Press.
Uimonen, P. 2020a. #MeToo in Sweden: Museum Collections, Digital Archiving and Hashtag Visuality. *Ethnos. Journal of Anthropology* 85 (5): 920–937.

———. 2020b. *Invoking Flora Nwapa. Nigerian Women Writers, Femininity and Spirituality in World Literature*. Stockholm: Stockholm University Press. Open access: https://www.stockholmuniversitypress.se/site/books/m/10.16993/bbe/.

Westermann, A. 2022. Against the Aestheticization of Technofossils. Considering Migrant Labor and Petrochemical Feedstocks in the Future History of Plastics. In *Anthropocene Markers: Stratigraphy and Context, Anthropocene Curriculum*, ed. C. Rosol and G. Rispoli. Berlin: Max Planck Institute for the History of Science.

Wright, T. 2008. *Visual Impact: Culture and the Meaning of Images*. Oxford: Berg.

CHAPTER 8

Trash Diving: A Global Comparison—An Expanding Rena Mälaren and Trash Divers in Brazil and Thailand

The discipline of social anthropology has traditionally been one of comparison and contextualization. In this regard, the following is a comparison of the work of Rena Mälaren in Sweden and that of trash scuba divers in Rio de Janeiro and Arraial do Cabo, Brazil, and on Koh Samui, Thailand. We will examine the similarities and differences in the socio-political situation within which these groups operate, and survey their varying solutions to context-bounded situations. Not only are the situations on land different, but their underwater worlds differ in composition, with implications for trash diving practices. In Thailand we will further engage with a land-based group that collects marine debris that has washed ashore. Rena Mälaren collaborates with similar organizations in Sweden, and therefore I find this comparison important.

Initially we will follow Rena Mälaren to two Swedish cities beyond Stockholm, that of Västerås and Gothenburg. The trash scuba diving group seeks to expand in both cities through partnerships with state bodies as well as in other localities. In Västerås we learn about how Rena Mälaren began a collaboration with the local fire department, hoping that other departments around Sweden will follow suit. Gothenburg was the first stop on the group's initial trip outside of Stockholm with the purpose of expanding operations and inspiring local divers in other cities to start doing similar work. We then consider environmental ethics as a means for

R. Rodineliussen, *Underwater Worlds*, Sustainable Development Goals Series, https://doi.org/10.1007/978-3-031-63370-6_8

social movements to make change happen. The international NGO Project AWARE provides the context for the case of Brazil and Thailand. Then, I present another organization called Trash Hero, and we join them on a beach cleanup, this time also in Thailand. Finally, we point out the similarities and differences between the case of Rena Mälaren and those in Brazil and Thailand.

Rena Mälaren: Expanding

> For the time being, I think Rena Mälaren has expanded as much as it is possible within its original format and geographical area. To continue the expansion, to have a growing impact on environmental politics in Sweden and increase the amounts of waste removed from water, we need to reorganize somewhat. Mainly: expand to other cities in Sweden, and then internationally. With our increasing funding we will be able to help fund cleanup work in other localities and assist them in a way that will get their work going quickly and easily. New local groups will not need to go through the economic problems we did. Instead, they can start to clean right away!

The above was said by Fredrik during a discussion I had with him in early 2022 when in the midst of planning ways to reorganize Rena Mälaren and expand operations outside of Stockholm. He had said similar things on many earlier occasions, but now the planning began to be transformed into action.

Early in 2022, Rena Mälaren visited the Swedish towns of Västerås, Strängnäs, Uppsala, and Gothenburg. Below, I elaborate on the visits to Västerås and Gothenburg. These two locations are particularly good to think with. This is because Västerås is an example of a low-visibility dive that created a collaborative relationship between Rena Mälaren and a local fire department. Gothenburg was the beginning of "Dykare Holgerssons Underbara Skräp-resa" (Diver Holgersson's Wonderful Trash Journey), the official start of Rena Mälaren's expansion campaign to launch similar cleanup work among local divers across Sweden. A banner, shown below, that was made and the name of the campaign are inspired by Selma Lagerlöf's 1906 classic book *Nils Holgersson's Wonderful Journey Through Sweden* (Nils Holgerssons underbara resa genom Sverige) (Fig. 8.1).

In this book, the leprechaun Nils travels around Sweden on the back of a goose, experiencing many adventures. Here Rena Mälaren continues to

Fig. 8.1 The banner shows a diver on the back of a goose together with planned dates when Rena Mälaren will go to new towns. (Creator: Tone Lindholm)

draw upon familiar Swedish symbols and stories to engage the Swedish population by depicting Nils in a dive suit on the back of a goose, a strategy similar the Gävlebocken and the City Hall trash replica. Like Nils, Rena Mälaren will travel around Sweden, experiencing different adventures along the bottoms of lakes, rivers, and coastal areas near the towns visited. The group hopes that local diver and rope puller recruits will rally to the dives and continue the work after Rena Mälaren leaves.

Västerås: Collaborating with the Fire Department

Rena Mälaren had long been trying to establish a collaboration with the various fire departments in Stockholm, but without success. However, in spring 2021 a fire department in Västerås heard about Rena Mälaren's work and was interested. The Västerås Fire Department invited Rena Mälaren to join them in arranging a dive together. Once this collaboration became known, some of the departments in Stockholm also became interested in working together.

The fire departments divers are required to have "water time" regularly, that is, training to retrieve objects in the water in order to be ready, should they have to search for a human body. The idea proposed by Rena Mälaren, and that the Västerås Fire Department agreed to, was that since these divers had to be in the water searching for objects anyway, it would be better if they searched for waste. This would combine their training with doing something good for the environment.

On an early spring morning in 2021, the group from Rena Mälaren arrived in Västerås, a town located 106 kilometers to the west of Stockholm, along Lake Mälaren. The members were warmly greeted by the fire department. As everyone was eager to get into the water, conversations were brief. The fire department divers appeared confident as they wished the divers from Stockholm best of luck while everyone geared up and got into the water. It quickly became apparent that conducting trash dives in Västerås would in some ways be even more challenging than in Stockholm. Because Västerås is located further into Lake Mälaren, there is less underwater visibility due to the increasing "brownification" of the water. This process is a common phenomenon in lakes when soil particles leak into the water and make it browner. In combination with high levels of eutrophication, this results in extremely low visibility. In the case of Västerås, the visibility was not just low, it was non-existent. In the vignette in this **book**'s introductory chapter, I describe the sensuous experience of diving in the water around Västerås. To give an example: when I wanted to test the level of visibility when I was in the water, I moved my hand slowly toward my eyes to see how close it would come before I could see it. I hit myself in the face without ever seeing my hand. Thus, the visibility was zero.

The equipment used by the fire department and the divers of Rena Mälaren differed somewhat. All wore drysuits, but while the divers from Rena Mälaren used regular goggles and an open breathing system, the fire department's divers had a full mask with communication to the surface, and their air supply was connected to a tank on land, providing them with an unlimited supply of air.[1] The fire department's diver was therefore linked to the surface by an air hose and a rope attached to it. They then used it similarly to Rena Mälaren's ropes connecting diver and rope puller. What became evident during the dive was that experience and regular water time make a great difference. The divers from Rena Mälaren were more efficient than the fire departments divers in locating waste in the water and in navigating without sight, even if they had less advanced equipment. If the fire department's divers were self-assured before the dive, they were much humbler afterward and keen to share experiences and tips as to how to better move around without being able to see, and how to communicate between diver and rope puller.

The dive was a success and volunteers collected about three tons of waste. The fire department immediately asked Rena Mälaren if they would like to come back again. It was decided that the group would return to Västerås about a month later. In the meantime, the fire department would have more time to practice, and improve their techniques. Moreover, the fire department wanted to make the next dive into a media event to advertise this new service they would be providing for the public in Västerås. Even this first dive gave the fire department in Västerås some publicity after they posted pictures of it on their social media. They soon were "tagged" on Instagram by another fire department, in Wolfsburg, Germany, with a picture of that department conducting a trash dive in a local lake after being inspired by the work of their colleagues in Västerås. This triggered a lot of enthusiasm within the local fire department, and they became even more eager to keep up trash diving as part of their training routine.

The trip to Västerås was successful in creating a collaborative relationship with a fire department and getting them to use their water time for

[1] To invest in full face-masks or not has been a recurring question among the divers of Rena Mälaren. They have concluded that although full masks are useful in that they provide a wider field of vision, they are less safe than regular goggles when conducting trash dives. This is, because if one gets caught in something or runs out of air and another diver needs to assist you with a spare octopus to share his/her air with you. Then, if you have a full mask, you must remove the whole mask, making it impossible for you to see anything at all. The fire department's divers on the other hand, are connected to land by a hose with an unlimited supply of air, so for them full masks can increase safety. But, if you dive with tubes, they do not. Therefore, the group has opted not to invest in full masks.

trash diving. Other departments were inspired to do the same, as was the case with the Wolfsburg Fire Department. As Fredrik happily told me after the trip to Västerås,

> If we can pursue fire departments to do trash dives as part of their training, then we will get a lot of waste out of the water, and it will be during work hours. It is great to have all our volunteers, but if volunteer power can be combined with the work of paid divers, then we will see a lot happening!

Gothenburg

Rena Mälaren dived in Gothenburg on Saturday and Sunday, 26–27 March 2022, the first stop on Dykare Holgersson's Trash Journey to spread awareness and inspire local divers to start trash diving. The Saturday dive was along the canal in the City Center. Since the weather was fine with blue skies, sun, and 15 degrees Celsius, it looked like the whole town was walking along the canal, watching how objects came out of the water. Most of the crowd looked truly amazed, and even a bit disgusted over all the things being pulled up in front of them while they ate their lunch or enjoyed a beer in the sun. Many had not heard about Rena Mälaren before, but some had. One older man approached me with a smile while I and some of the divers geared up to get in the water. He asked, "Do you know that there is a similar project in Stockholm? They also dive for trash!" The man looked disappointed and amazed at the same time when he realized that it was the Stockholm group that was preparing to dive in front of him. He was disappointed because he had hoped locals were doing it. But he was amazed that people had traveled 450 kilometers from Stockholm to clean up Gothenburg. The encounter showed that although Rena Mälaren is mostly known in and around Stockholm, their reputation has begun to reach further (Fig. 8.2).

There were 30 in the group that had traveled together by bus from Stockholm to Gothenburg for this event. Due to the relatively large number of divers and rope pullers, much waste could be collected. For example, on the Saturday, Rena Mälaren reached a milestone of 1000 electric scooters pulled from the water. The group needed to find 34 scooters during the visit to Gothenburg to reach this goal, and they found 39, along with 1.5 tons of other waste. The visibility in the canal was about 10 centimeters. Thus, it was a diving experience similar to the one described in Västerås. The divers mainly had to crawl around on the bottoms,

Fig. 8.2 Left: One of the divers doing some "freediving," and then showing off with a shimmering hat he found in the canal. (Photo: Author); Right: Onlookers, rope pullers, and a diver in the central canal, Gothenburg. (Photo: Joakim Hall)

searching for waste by moving their hands in front of them. This was another factor in the onlookers' astonishment, for, as I was told several times, they could not understand how anyone would voluntarily submerge themselves in that brownish water. The Rena Mälaren members from Stockholm were joined by three divers from a local diving club, who wanted to begin trash dives on their own. They joined Rena Mälaren for the entire day, learned communication techniques between divers and rope pullers, and got to experience low visibility diving in a way they never had before. The three local divers were all very experienced and moved with confidence in the water, even though diving under these circumstances was new to them.

After the first dive of the day, the local divers sat on the shoreline in the sun, still wearing their drysuits, and discussed the conditions they had encountered in the water with the divers from Rena Mälaren. The local divers wanted to know if this was the "normal" condition of the waters that Rena Mälaren dives in, or if this was worse than normal. When the answer was that this was normal for the group, the local divers shrugged their shoulders while posing sad smiles. "Well," one of them said, "I think I can get used to it, I guess. It is not nice, but it is also not bad—it is what it is." (Fig. 8.3).

For the Sunday dive, the group went to the Gothenburg Royal Sailor Association's (GKSS) marina. Here a similar situation to the one with the

Fig. 8.3 Some of the 39 electric scooters collected during the Saturday dive in Gothenburg. (Photo: Joakim Hall)

older man occurred. After the group had been working for about an hour, many objects had been connected to lines in the water. I had dived for the hour before coming up to assist in the pulling part of the cleanup, as many of the objects were bulky and heavy. Then a local man from the marina approached and said, "Why did they not ask us to do this cleaning? I mean, our dive boat is just over there. We are a local diving club." But then he looked a bit ashamed and continued, "Well, it could of course be because we have not proposed it ourselves, as I assume you did." He then joined in pulling ropes, carrying waste, and doing his best to be a part of the effort. The confusion, almost irritation first shown on the man's face quickly changed to determination to join in the cleanup. He talked about the possibility of doing similar dives to keep the marina clean in the future and was visibly inspired.

The water at the marina was clear, with visibility at close to 4 meters. As a result, waste was mainly found further out along the bridges, where it could not be seen from above—emphasizing the point I make that water is a sink for waste when it cannot be seen. In addition to good visibility, the bottom was full of life, making it necessary to look into each piece of trash brought up to remove crabs, fish, and starfish and return them to the water (Fig. 8.4).

The visit to Gothenburg was successful in that local divers worked alongside Rena Mälaren and experienced what it means to do trash dives with little visibility. Rena Mälaren were successful in getting the machinery of the authorities involved and they removed the waste collected after the dive. Local divers could maintain this contact and continue the work. Then at the GKSS marina, another group of local divers became interested and involved, although on land, in cleaning up their water. Thus, the goals of Rena Mälaren were achieved and the campaign moved on to the next town.

Fig. 8.4 The author pulling up a large metal container. (Photo: Joakim Hall)

To Make That Environmental Ethic Stick

Ethics and morals are two key topics of discussion both within Rena Mälaren, and between its members and the public and authorities.[2] These discussions take for granted that no trash should end up in the water and that throwing trash in the water is morally wrong. Such ideas are embodied by members in making "their bodies battlegrounds for environmental and social change" (Mathias 2020: 257) when submerging themselves in cold and often visibly dirty water. They document where the waste is now lying, show the complications and filth involved in bringing it to the surface, and argue for a change in how waste is managed and how water is treated. The group believes that it is by *showing* rather than only *telling* about the issue that one can change the opinion and habits of the public and officials. The hope is that individuals who have seen the work of Rena Mälaren will become conscientious about not contributing to the problem of waste in the water. The group has thus made this an ethical issue (Kleinman 2006). Rena Mälaren seeks to utilize the relational and reactional capacity of emotions by attaching an emotion of disgust or anger to waste objects in the water among those who observe the cleanup work (Ahmed 2014: 8). This is done with the hope that the next time a person acts in a way to pollute the water, that person will hopefully remember this feeling of disgust/anger and avoid such an outcome by acting differently. Thus, preventing additional waste from ending up in the water.

To exemplify how this ethical standpoint is shared with authorities and how it can be made to "stick" (to use John Mathias's (2020) terminology about "sticky ethics"), let us recall the example of dirty snow being dumped in Lake Mälaren. Here Rena Mälaren had to create an event that would attract an audience interested in listening to the message the group wanted to share, for, as Michael Warner (2002: 61) writes, "A public exists only by virtue of address, it must predicate some degree of attention."

[2] In recent years, the terms ethics and morality have been increasingly applied in the anthropological literature (cf. Fassin 2012; Zigon 2008). However, as Julia L. Cassaniti and Jacob R. Hickman (2014) point out, this can be a problem because numerous authors write about ethics and morals in different ways, with different meanings and assumptions, and still use the same term as if it had a clear meaning. While this is beyond the scope of this book, the issue should be acknowledged. In this book I think of ethics as a "moral" principle guiding an individual's actions that is based on moral assumptions on what is right and wrong. Hence, I do not make claims that one person's ethical or moral principles necessarily are the same as those of other people.

Rena Mälaren did this by staging an underwater demonstration deriding the government's dumping of dirty snow in the water, which they livestreamed, filmed, photographed, and then shared online on social media along with their message about the authority's polluting actions. When Rena Mälaren accused the authorities of Stockholm of a double standard in their announced concern for safeguarding the environment while dumping toxic waste in the water, the group was trying to make this ethical position stick. It now became difficult for the authorities to dump snow and still claim their full support for a clean marine environment. Although the dumping continues, this resulted in growing public awareness of the issue and a build-up of pressure to abandon the practice. When Rena Mälaren accused the authorities of negligence, the authorities tried to justify their actions. In the end, it seems like the authorities were seen as accountable for the additional waste dumped in the water. This public opinion can be gauged from the comments on Rena Mälaren's Instagram account, in reports on TV, and in comments made to members, including myself, present at cleanups. On a number of occasions, people walking by have voiced their concern and frustration about the way the authorities dispose of the snow, and thanked members of Rena Mälaren for pointing this out and making this practice visible. Rena Mälaren aims to continue pressuring the authorities on this matter, building on the momentum created by their first underwater demonstration. When Rena Mälaren voices ethical and moral claims like the one about the authorities dumping dirty snow in the water, they make what we have called aquabiopolitical relations visible by highlighting the damaging and violent relationship between humans, waste, and water.

Environmental Activism Compared

Project AWARE is an environmental scuba diving non-governmental organization (NGO) created in 1989 by the Professional Association of Diving Instructors (PADI), the world's largest scuba diving association. It is present in about 180 countries worldwide. In 1992 Project AWARE became independent of PADI, although still closely connected to the mother organization. Project AWARE was first recognized as an NGO by the US in 1992, in 1999 by the UK, and in 2002 by Australia. However, in 2021 Project AWARE rejoined PADI and changed its name to PADI AWARE. The motivation for this move is, according to the organization's statement on its website,

> PADI's strengths in numbers, a network of over 128,000 PADI Professionals across the globe, combined with AWARE's strengths in conservation, 30 years of program deployment and policy expertise, make a truly unique voice for the protection of the underwater world. Acknowledging that we are stronger together, we are committed to our joint vision for a balance between ocean and humanity. (https://www.padi.com/aware/history, accessed March 16, 2022)

The organization has two main goals: the preservation of rays and sharks and removing marine debris. It is the latter that mainly concerns us here. Although the organization changed its name in 2021 from Project AWARE to PADI AWARE, we will refer to it as Project AWARE since that was the name that applied during my fieldwork in Brazil in 2017 and Thailand in 2019, while working with divers associated with or inspired by the organization. There are several similarities between the work Project AWARE and Rena Mälaren do. For example, both organizations tell a similar story of how waste ends up in the water and both emphasize that as long as trash is below the surface, most humans forget about it. On the former Project AWARE website (accessed in March 2017) it was stated:

> By visualizing this information, we're revolutionizing the way the underwater marine debris issue is portrayed—sharing an accurate perspective about a threat that has previously been disregarded as out of sight, out of mind.

This statement aligns closely with the work of Rena Mälaren, and the case in Gothenburg where waste was only dumped out in deep water. Project AWARE has also created an infographic about what they call "The Ugly Journey of Our Trash." It details how trash moves from society and fisheries into the environment and eventually ends up in the water, where it has harmful effects on marine life. The infographic was also posted on their earlier website and is therefore no longer available.

However, there are also major differences between Rena Mälaren and Project AWARE, mainly in the type of waste collected, the type of visibility of the water they dive in, as well as branding and operating for profit. If we look at a diagram produced by Project AWARE showing what kind of waste they collected in 2020 (link in footnote),[3] we see that it is mostly small items and plastics, contrasting considerably with what Rena Mälaren

[3] https://issuu.com/projectaware/docs/pa-2020-report-digital, accessed 19 April 2023.

finds around Stockholm. The reason for this difference has much to do with visibility, as we shall see below.

With regard to finances, Rena Mälaren is committed to expanding its efforts by funding other groups to do cleanup work. Project AWARE, on the other hand, requires payment from its members to allow them to be affiliated with the organization. These similarities and differences raise the question of how Rena Mälaren can expand to the extent the group hopes without assuming the more business-like qualities of Project AWARE. To answer this question, we need to look into how divers engaged with or inspired by Project AWARE work for a cleaner marine environment. We begin this journey with a visit to Rio de Janeiro and Arraial do Cabo, Brazil, then traveling to the Thai island of Koh Samui.

Project AWARE in Brazil: Rio de Janeiro and Arraial do Cabo

I was in Brazil for almost three months in the winter of 2017 to conduct fieldwork for my master's thesis in the Department of Social Anthropology at Stockholm University (Rodineliussen 2017). Before leaving Sweden for Brazil, I was already talking to Luiz, who would become one of my main interlocutors. Luiz, a man in his 30s, lived in one of the fishermen's colonies along the coast of Guanabara Bay in Rio de Janeiro. Luiz is from a family of politicians. He is in the army, where he earns his living, and works as a scuba diving instructor in his free time to make ends meet. As a diving instructor, he is in liaison with PADI and Project AWARE and does his best to work for the organizations and implement their values in his daily life. As Luiz put it, "Project AWARE provides the tools that I use to change the reality of my life, which is Rio de Janeiro." Luiz goes to great lengths to encourage other divers to be part of cleanups around Rio, and he even offers free scuba diving lessons for those who participate with him in more than two cleanups. Luiz also collaborates with a youth prison center close to his home. Youths at the center join him in cleanups of the mangrove forest in the area as part of their rehabilitation into society. Luiz hopes that this type of work might impact the way these youths engage with the environment in the future, emphasizing that it is the younger generation that is the key to a future where waste will not be thrown into the environment as it is today.

The work of Luiz is similar to that of Rena Mälaren in that he focuses on cleaning trash from the water. But even more important to him is to share awareness, make individuals stop littering, and end practices that increase the amounts of waste entering the water. Guanabara Bay, the body of water along Rio de Janeiro, is extremely polluted, and levels continue to rise. The pollution was depicted in an article by John Branch in the *New York Times* in the wake of the Rio Summer Olympics in 2016, mentioning that calls had been made to move the venue for water sports due to the high levels of pollution in the bay (Branch, NYT, 27 July 2016).

In this article, the journalist asks Paulo Rosman, Professor of Oceanography and Coastal Engineering at the Federal University of Rio de Janeiro, about his view on the increasing pollution. "O.K., stop. Turn 180 degrees. That's the problem. Not the bay. The bay is just a consequence." This statement is followed by locals describing how rivers leading to Guanabara Bay are viewed: "The children don't know it as a river. They know it as a sewage trench. What arrives in the bay is produced here." Furthermore, the article cites promises that were made to clean up the bay prior to the Olympic Games—promises that never were kept. When I asked Luiz about those promises, he just laughed ironically, saying,

> Those were just empty promises, of course. Money was never allocated for cleaning the bay; it was put directly into constructing things needed for the games. These structures then leaked further polluting materials, mainly a lot of salt, into the water. This has just increased the harm done to the marine environment, especially the mangrove forest that is so important as a nesting habitat for small fish and birds.

To Luiz, it was never realistic to trust the politicians to act on behalf of the environment. This was something individuals like himself had to do on their own if something were to actually happen. The idea among local children that the river is a sewage trench rather than a body of water is a key part of how Luiz understands the issue, and why he works with the youth. He is trying to change their perception of the environment, make them think of the future and understand the difference between a living and a dead Guanabara Bay. The story told in the abovementioned article is similar to Project AWARE's "The Ugly Journey of Our Trash," as the river moves past villages, collecting waste and leaving it in Guanabara Bay.

Luiz is vocal in his attempts to create an emotional reaction among locals to change their behavior and begin to clean the environment around them. He does this by, for example, organizing informational meetings and cleanup actions where he first collects waste from the water and then takes pictures, posing next to the waste. These images are then posted on social media. The waste is weighted and Luiz reports the number of items to Project AWARE's system as part of the organization's citizen-science project to map levels and currents of waste in water globally.[4] These images are important for several reasons. They assist Luiz in providing proof of cleanups that he can show locals during the meetings he organizes to demonstrate that he is not only talking but actually engaged in cleaning.[5] Then, when these images are posted online with the hashtag of Project AWARE, they place Luiz on the Project AWARE map as an environmentally aware diving instructor and as a member of Project AWARE.

Project AWARE has an interactive map on their website that shows their affiliated diving instructors and centers. To appear on this map is important for Luiz as it acknowledges that he is doing the work promoted by Project AWARE. It also helps attract the attention of the growing "eco-diving" community. "Eco-divers" represent a growing sub-group of divers who seek to explore the undersea world, but to do so in an environmentally friendly and sustainable way.[6] Thus, being visible online and associated with Project AWARE has two-fold importance for Luiz. It helps him in his ambition to share his awareness of the importance of cleaning the marine environment, and it insures a larger number of diving-tourists finding their way to him, which in turn allows him to further influence them to engage in cleaning while diving.

Prior to my visit, Luiz suggested that I should visit the nearby town of Arraial do Cabo, renowned for its beautiful beaches and scenery, if I wanted to see how being visible is important to attract certain tourists and what visibility allows a dive center to do with regard to cleaning the environment in a locality fraught with different political and economic interests. Arraial do Cabo is a fishing town with about 27,000 inhabitants and is located in the Regiao de Lagoa, approximately 200 kilometers along the coast from Rio de Janeiro. The town is unofficially known as the "diving capital of Brazil," according to first Luiz, but then also other divers

[4]" Citizen-science" here refers to a research project conducted mainly by lay people without scientific training.

[5]For a discussion on images as proof see Chap. 7.

[6]For an overview of studies on scuba diving tourism, and its possible environmental impacts, see Musa and Dimmock (2012, 2013).

concurred. The reason behind this popularity among divers is the rich marine life, which is a result of the uncommon phenomenon called an "upwelling" that occurs in the area. This means that warm surface water is blown away by winds, allowing nutritious cold water to reach the surface and enrich the water. In 1997 the Brazilian government declared the area an extractive marine reserve to prevent overfishing and preserve traditional fishing families and their customary practices. Luiz told me that one diving center in particular in Arraial conducted regular cleanups. While not officially associated with Project AWARE, the center is inspired by it and does the same type of work, branding themselves environmentally aware. The reason they are not associated with Project AWARE is that they are unwilling to pay the membership fees needed for this association. The owner, whom I call Ella, is a woman in her 40s who has worked with diving in Arraial for over 20 years (Fig. 8.5).

Fig. 8.5 The street leading to Prainha Beach, Arraial do Cabo. Showing waste littering on the street, ready to be swept toward the water by winds or rain. (Photo: Author)

When I arrived in Arraial, I walked around the town, visited the different diving centers, of which there were 10 at the time, and walked along the beaches observing the surroundings while trying to get a sense of the town. I noticed that waste littered street corners, as in the image above, collecting in the bushes, on the beaches, and floating around in the water. When I walked along the beaches, I usually kept a bag in my hand, collecting the waste I found along the way. In the image below, you find what I picked up from cleaning an area of 2 × 2 meters selected at random along one of the beaches. The bag contains scraps of fishing net, plastic bottles, aluminum cans, and a mixture of hard to distinguish bits and pieces of plastics. The situation was similar on the other beaches, and I collected numerous bags-full like the one below during my time in Arraial (Fig. 8.6).

Fig. 8.6 Waste collected from the beach at Praia dos Anjos. (Photo: Author)

Arraial do Cabo has one main street connecting the western and eastern beaches. The pier is located at the eastern beach with the name Praia dos Anjos. This is the busiest part of town. At the time of my visit, the pier had only three mooring lines, with many boats competing for space. Fishermen, tour boats, diving boats, and private vessels all wanted access to the pier, especially in the early morning and late afternoon (Fig. 8.7).

What played out around the pier can best be described as a theatrical play of boats rhythmically moving back and forth as numerous fishermen, diving boats, and tour operators shared the same slots along the pier to load and unload passengers and catches of fish (Rodineliussen 2019). The small pier, the large number of fishermen, diving, and tour operators, and the many tourists are why conducting cleanups in the water around Arraial is complicated: the location with the most significant amounts of waste in the water is also where there is the most boat traffic, namely, around the harbor. To clean this area, the diving center must obtain permission from the local authorities, who have to then close the area to boat traffic for a few hours. This is difficult because of the already crowded schedules among the different boats and the importance of the pier for the town's economy. As Ella explained to me,

Fig. 8.7 Boats parked in the vicinity of the pier at Praia dos Anjos. (Photo: Author)

> Neither fishermen nor tour operators are keen to close the water for traffic since it will impact their business. It is already so hectic to figure out when and who should dock at the pier. When we close down the harbor for traffic, this puzzle becomes even worse, and, since the town relies on tourists it is hard to convince politicians to bar boats from the water, even for a few hours.

Ella pointed out the waste that had been thrown around on the beach, which, according to her, was done by tourists. She observed that if tourists care this little about visible waste on land, why would they care about the waste below the water? This was of considerable concern to Ella, as her main argument when asking officials for permits was that she helped keep the water and beaches clean, which would have a positive effect on tourism. But if the tourists cared little about the waste, the authorities apparently cared even less. Thus, Ella asked her diving instructors, and those diving tourists who were willing, to clean the water as they dived. The waste collected was then photographed and posted on social media to attract further environmentally concerned divers. Ella argued that if she could interest enough tourists and become an important economic actor in Arraial, she would be in a stronger position regarding permits. I asked if she would not gain momentum by joining Project AWARE and utilize the visibility they provide. She answered that in a way she would, but she simply could not afford it. Her finances were tight as it was, and she was unable to send percentages to Project AWARE for work she could do herself, especially since her primary motivation was to clean waste from the water, not market her company.

Ella's reluctance could also be due to Project AWARE‘s minor presence in Brazil. Since the majority of tourists coming to Arraial are domestic, this liaison would not garner enough extra tourists to be worth the cost. An online presence, showing cleanups and raising awareness about waste in water, attracted enough of the eco-tourist divers as it was. And therefore, Ella could not see the reason to join the organization. Luiz, on the other hand, based in Rio de Janeiro, mainly dealt with international tourists, many of whom knew of his liaison with Project AWARE. Thus, the relationship was well worthwhile for him, even if it came with a cost.

Project AWARE in Thailand: Koh Samui

I traveled to the Thai island of Koh Samui for fieldwork at a dive center affiliated with Project AWARE at the end of 2019 and stayed for a month. I had previously gone to Thailand as a scuba diver on numerous occasions for over a decade. For example, I had taken my Advanced Open Water certificate (allowing you to dive as deep as 30 meters) and conducted over 40 dives in Thailand prior to my doctoral fieldwork. Thus, the dive scene in Thailand was familiar to me, and this longer perspective informed my analysis. I went to Koh Samui because it is one of the main tourist destinations in Thailand, and the island has a significant number of diving centers. It also shares diving sites with the renowned "diving island," Koh Tao. When I was not below water or in the diving shop, I was also affiliated with the local Trash Hero group, cleaning beaches around the island.

In collaboration with PADI, Project AWARE has developed a specialty diving course named "Dive Against Debris." In my conversation with the Project AWARE affiliated diving center on Koh Samui prior to my visit, I had asked about this course. I decided to take this certificate as part of my engagement with the center. The course entailed a two-dive adventure with an instructor, theoretical and practical readings about why it is essential to remove marine debris, how cleaning should be done so as not to affect the environment negatively and as a last step, logging the waste collected into the Project AWARE system to contribute toward a more informed understanding of how much waste there is in the water across the globe, and how waste moves when in water. One annual count from this system is shown in the diagram discussed above called "Debris Items Breakdown." The waste found on dives around Koh Samui corresponds to the diagram, with "plastic fragments," "plastic food wrappers," and "plastic bags" being the most common items found in the water.

When I arrived at the diving center for the first time, it was early in the morning. As the door opened, I was greeted by the familiar smell of yellow label Lipton tea, powdered coffee, and rubber—the same smell you will encounter at almost any diving center.[7] I had not been inside one for some time and the smell hit me hard, bringing back memories of earlier diving experiences in Thailand and elsewhere. Thus, with a smile on my face and

[7]This conclusion I draw from my own experiences from various dive centers in: Sweden, Denmark, Spain, Greece, Malta, Thailand, Vietnam, Cambodia, Malaysia, Israel, Zanzibar, and Brazil.

a pleasant feeling inside, I walked through the door, where I immediately noticed two posters. One asks the viewer to "Say No to Plastic," and the other advertises Project AWARE. I stopped to look at the posters before presenting myself at the counter. I had a brief conversation with the owner, a woman in her 40s from Germany, before the instructor I had talked to about the specialty course appeared.

My instructor, I will call him Jamey, was a man from France in his late 40s, walked me around the diving center, showing me the assembled gear and images of diving sites. He talked about his engagement with underwater cleanups, his connection to Project AWARE, and the specialized diving course we would be doing together on the following day. Jamey also told me his story and how he ended up here on Koh Samui as a diving instructor. He said he had always known that what he wanted to work with was scuba diving, so he began as a seasonal divemaster (the first professional level of diving), when he was in his early 20s. The job meant that he had to move with the diving season, working a few months at one diving center, then going on to another.[8] During this time, he became acquainted with Project AWARE after noticing how much waste there was in the water and that the amount seemed to increase year by year. He thus started to seek to work at diving centers affiliated with Project AWARE to help promote the idea of "Dive Against Debris": that divers should collect the waste they encounter on all dives. According to Project AWARE, not all divers need to conduct regular cleanups, but if they bring the waste they encounter on dives up with them, much waste will be collected, with minimal effort required by the diving community. Jamey also wanted to "adopt a dive site," another idea of Project AWARE, meaning that one diving site is repeatedly cleaned from waste and monitored to see how quickly new waste reappears at the cleaned site. Tracking this allows an understanding of the movements of waste in water and waste into the water from land-based sources in the area.

The Specialty Diver Dive Against Debris course made evident similarities and differences between trash diving in Thailand and Sweden. The two most notable differences were temperature and visibility. In Sweden, the water temperature for most of the year is between 2 and 10 degrees

[8] Being a seasonal worker is common to scuba divers in their early career. Most people who end up working as dive professionals start out as scuba enthusiasts, then search for work that will allow them to combine pleasure with a paying job. For this reason, you can meet instructors and dive masters from almost anywhere in the world at tourist destination diving centers.

Celsius, and in Thailand, it is around 25 degrees Celsius. Visibility in the water around Stockholm varies from half a meter to two meters, at best. Around Koh Samui, the visibility varies between 15 and 40 meters. These two differences require different gear (drysuits for Sweden and wetsuits in Thailand), which means that divers in colder water have more things that can be damaged during a dive and that they can also move less freely in the water. Higher visibility allows a diver to easily spot trash dangerous items to avoid, allowing more effective planning of the dive. Another difference is the type of waste that is present. In the waters around Stockholm, divers find big pieces of scrap metal, car batteries, and other large items. Around Koh Samui, the items found are mostly smaller, such as pieces of plastics, rubber, or metal cans. This is, however, not so much a difference between Thailand and Sweden as it is a difference between urban and island trash diving. The trash in the waters of Stockholm can be sharp and toxic, which is less the case around Koh Samui. However, in Thai waters there is another potential danger, if divers are not careful when picking around the waste. They can, by mistake, come in contact with a stingray, sea urchin, or a scorpion-fish lurking around the waste, which could be a painful or even dangerous experience for the diver.

The similarities between Stockholm and Koh Samui in trash diving were mainly practical ones. A diver first searches the bottom for waste and then picks it up. In Koh Samui they often collect the waste in a mesh bag hanging from the diver's belt. The mesh bag allows water to pass through it and will, therefore, not become heavier than the items placed within it. Mesh bags are, however, rarely used in dark water as they easily can get stuck on objects on the bottom that the diver cannot see. Like the divers in Stockholm, Jamey carefully selected what items to take from the bottom to ensure that he did not remove things that had become habitats for marine life, unless they were toxic or otherwise dangerous to marine organisms. Bottles made of glass are an example. These can effectively be recycled, saving energy and resources. But they can also often be inhabited by small crabs who have adopted them as shelters, enjoying the safety provided by the glass. Moreover, since the bottles are not toxic, they are sometimes better left on the bottom for continuous use by, in this case, crabs (trash as a habitat for marine life has been considered in Chaps. 4 and 6). However, bottles can also become a trap for the crabs since, if they move in when they are small, they might be trapped once they grow bigger.

After I had completed the specialty course, I continued hanging around the center, talking to visiting divers, and joined diving excursions to see how the center implemented its environmental thinking when providing so-called fun dives for tourists. The visiting divers where mainly western tourists from countries such as Sweden, Germany, France, the UK, and the US. They were evenly divided between males and females, often in the ages between 20 and 50, and had in common that they were middle or upper-class. Scuba diving is an expensive hobby, and therefore mainly individuals with sufficient economic means can take part in this activity. The image above is from one of these fun dives, with another instructor from the center taking a picture of one of the tourists removing waste found along the bottom (Fig. 8.8). Before each dive, there was a briefing about

Fig. 8.8 Scuba divers finding trash during a recreational dive, following the proposal of Project AWARE to make all dives a Dive Against Debris. (Photo: Author)

the diving site, safety, and rules of conduct. At this briefing, the instructors from the center emphasized the importance of not leaving any trash on the boat so that it could be swept away by the wind and end up in the water. Furthermore, the divers were asked to keep an eye open for trash during the dive and, if they spotted any, to bring it up to the boat. After one of these dives, on the way back to the center, I sat in the front of the boat with one of the instructors while he explained to me and some of the other divers how the center continuously cleans the diving sites in the area. They go there at least once a month and clean them from waste. This is in addition to removing the waste they find during fun dives with tourists. In high season they do not have time for more cleanups elsewhere than at the regular diving sites, but during the low season they clean other locations, too. This pattern of cleaning tourist diving sites is similar to the cleanups conducted by the diving center run by Ella in Arraial do Cabo, Brazil. The regular cleanups had the dual purpose of cleaning the environment and making the diving experience for tourists more pleasant. To keep the marine environment visibly clean is thus important for economic and environmental reasons.

On another day, I walked along the beach with Jamey, sporadically collecting waste as we went. He told me that he always carries a bag with him so that when small items of waste cross his path, he can collect them. This prompted me to ask if he knew of the work of the local branch of Trash Hero. Jamey answered that he knew of the "outstanding work they did cleaning the beaches around the island." But he felt that he already did his share for the environment with his dives against debris and this walk-and-collect of smaller pieces along the shore.

Trash Hero Koh Samui

> Trash Hero is an energetic, volunteer-led movement that drives change within communities around the world, motivating and supporting them to clean and prevent plastic waste.[9]

The quotation above is from the website of Trash Hero, an organization with affiliates in numerous countries, one of them being Thailand. The Thai Trash Hero affiliation is active on national and local levels, with

[9] https://trashhero.org accessed March 25, 2022.

smaller branches in various locations, such as the one I visited on Koh Samui.

Trash Hero's main goal is to engage local communities as "grassroots" affiliates. By uniting their efforts, locality to locality, they can scale up their impact to reach national and then a global level. All members use social media to document and showcase their work. This is how they gain followers, supporters, and new volunteers. Increased media attention is an essential tool in the quest to change how waste is managed and to make local politicians around the globe take waste management and cleaning the environment more seriously than they have in the past. The idea of becoming increasingly visible on social media to enlarge the group's sway over politicians is similar to how Rena Mälaren operates when posting images of waste removed from the water online. In Sweden, Rena Mälaren has initiated some collaborations with the state-funded organization *Håll Sverige Rent* (Keep Sweden Clean). In Chap. 6 I explain that this organization initially was hesitant to collaborate with Rena Mälaren, but as the divers got increasingly visible, this organization got interested in collaboration. Håll Sverige Rent organizes beach and park cleanups around Sweden, working much the same way as Trash Hero. Both organizations focus on waste that is yet not in water, or has come back up, and are therefore of importance for a broader understanding of how action can be taken against waste in the water and the larger environment.

My contact on Koh Samui was Ben, the local Trash Hero group leader. When I called him and told him about my intentions, he was welcoming and sounded curious to know more about my project. Ben is an expat from England in his 60s, who has been living on Koh Samui for several years. He views the island as his home and cares about its environment and the people and animals who live there. His group conducts cleanups every Sunday in different locations around the island. Ben had told me to come and join them this morning at one of the southern beaches of the island. The shore was lined with expensive resorts, claiming their private stretches of the beach. These areas of the sand were clean, without signs of waste. However, along the shoreline of the private beaches, waste piled up. Ben later said, with a heavy tone of disapproval, "If they are already cleaning, why not remove that waste, too? Instead, they let it stay on the public beaches that are right next to them." (Fig. 8.9).

When I arrived that Sunday morning, I saw a group all clad in yellow t-shirts, handing black plastic bags around while chit-chatting quietly. I

Fig. 8.9 Cleanup work in action. (Photo: Author)

heard occasional laughter among the chatter, and the air was filled with excitement. When I joined the group, I asked for Ben, and a tall man, brown from the sun, turned around with one of the warmest smiles I have ever seen. He took my hand, said welcome, put a black plastic bag in my other hand, and pointed toward the beach with a glimmer in his eyes, saying, "Let's talk after." And we began collecting waste. The work was conducted with bent backs, eyes on the sand, and hands moving from the ground to the bag and back, continuously picking up tiny pieces of plastics, rubber, bits of fishing nets, Styrofoam, bottles, cans, and many other unidentifiable items.[10] It did not take long before the first bag was full. I took another, and it soon was also full. I looked around, everyone was working efficiently, and the pile of bags grew by the minute, as seen in the image above. After two hours of intense picking, the beach was nowhere near clean, so Ben decided they would return the next Sunday and perhaps the week after.

The waste collected was left at a junction not far from the beach, where local authorities would pick it up. Ben told me that this was a relatively new arrangement. Not long ago, they had had considerable trouble getting the waste removed after they had collected it. But, as Ben explained, local authorities had become more willing to help, now that Trash Hero was so well-known. As noted earlier, things developed for Rena Mälaren similarly in Stockholm. Earlier, locals had tried to clean, just as Trash Hero did now, but without any response from the authorities regarding removing the waste afterward. Therefore, the local's actions had ceased. When Trash Hero first launched on Koh Samui, they had difficulties getting the authorities interested. However, although the authorities earlier did not care much about the group's occasional post on social media, they cared more now when the posts were regular and were viewed by an increasing number of followers. This, in combination with the national Thai government's drive at the end of 2019 to reduce the amount of plastics in the environment (by, e.g., banning certain types of plastics bags), had led to the local authorities on Koh Samui starting to help the group collect and sort the waste from the beaches.[11] As in Stockholm, growing visibility on social media proved effective in influencing the authorities to collaborate with the removal of collected waste.

[10] These unidentifiable items can tell many stories, and link with places (and times) from all over the world. For example, Anand Pandian (2021) calls them fossils of our time, writing "Messages in plastic were everywhere to be found within the cave, daring us to decode them." The stories Pandian found brought him from the island of Kefalonia to Germany. Michell Bastian and Thom van Dooren (2017) have labeled plastics the new immortals of our time. As they and Pandian in the paper cited above, stress, plastics endure through time, breaking into smaller and smaller fragments, but persisting none the less.

[11] This drive by the Thai government can be seen as part of the international campaign against marine debris, reflecting the S2S approach described in Chap. 4, whose aim is to eliminate potential waste items where they first appear (in the supermarket, for example), to not allow any chance of them passing through the waste management system and out into the environment.

After the cleanup I sat down with Ben, and he explained that the Trash Hero group meets once a week to clean one beach at a time around the island. Their aim is to encourage locals who tried to clean before without support from the authorities to begin again, this time with the aid of Trash Hero. As Ben said,

> I hope that in the near future we will mainly help the locals with their own cleanups by driving around the island and joining different villages on their cleanups. At this time, we are mostly expats who are involved, and this feels complicated. I would be more comfortable with locals leading this work, and us joining them, rather than the other way around. But as it has been shown, we have been more successful in getting the local authorities to collaborate. So maybe, I guess we, in a way, lay the groundwork for the locals to take over again. Or, so I hope.

Ben is aware that although he views Koh Samui as his home and has been living there for a number of years, he is still not *from* the island, and therefore he does not want to speak *for* the locals.[12]

What Can Be Learned from This Comparison?

From this comparison between the work of divers from Rena Mälaren in Sweden, divers associated with or inspired by Project AWARE in Brazil and Thailand, and beach cleaners from Trash Hero, a number of similarities and differences come to the fore.

Visibility on Social Media

Social media has become one of the most important ways for an organization to boost its visibility. This visibility has a dual purpose. First, in order for groups to be able to expand they need more volunteers or associates and they mainly get this by people finding them through online searches via social media. This is similar in Sweden, Thailand, and Brazil. What generally differs between these cases are the type of individuals the groups seek, and how long their engagement is needed. In the case of Rena

[12] I am aware that there is a considerable discussion going on about colonial practices, including well-meaning environmental work. This body of work is well-described and discussed by among others Max Liboiron (2021). The importance of this work, and these issues are undeniable, but they are outside the scope of this book.

Mälaren, new members are required for a long-term engagement. Luiz and Ella in Brazil, and the Project AWARE-affiliated dive center in Thailand are mainly seeking tourists to contact them for "fun dives" that will simultaneously result in cleaning the environment. As for Project AWARE itself, as an organization it solicits more paying associated divers and dive centers to fund its work, no matter where these divers are located. The only prerequisite is that they conduct trash dives in their locality and add to the Project AWARE database on marine debris. Trash Hero, on the other hand, seeks ongoing engagement, like Rena Mälaren, but is also happy for tourists to participate in one or more cleanups. It might look like the case with Rena Mälaren and Trash Hero are more utilitarian, for not looking toward tourists for finance, and rather see that these individuals engage in the work because they believe in the importance of it. However, Rena Mälaren and Trash Hero also utilizes their increasing visibility to attract support, although from organizations and companies, rather than individual divers paying for a diving experience. Thus, the difference is mostly *who* is paying and in what format, rather than the idea as such.

In all the cases presented in this chapter, there is an ongoing campaign to make environmental ethics stick (Mathias 2020), in that current and past waste management practices are exposed and publicized on social media for all to see. This is done with the hope that authorities and the public will be ashamed to continue with business as usual, and instead change the way waste and water are managed. This change has partially taken place because of Rena Mälaren in Sweden and Trash Hero in Thailand, where authorities have begun to help with the collection and sorting of the waste collected during cleanups. Thus, aquabiopolitical relations are evident, in how humans, waste, and water are collated in violent forms of relations where humans impose harm on life in the aquatic environment—harm that in the long run also harms humans when it "returns with the fish we eat," to recall the words of the UN Secretary-General's Special Envoy for the Ocean Peter Thompson whom we met in Chap. 4.

Visibility and Temperature in Water

One of the key differences between the Swedish case and those in Brazil and Thailand are visibility and temperature in the water. In Lake Mälaren, after removing a piece of waste from the bottom, a diver will see absolutely nothing. The water will be a blur of particles from the bottom mixing with the already low visibility of the lake resulting from eutrophication and brownification. In Brazil and Thailand, visibility on a bad day is about 10–15 meters, and up to 40 meters or more on good days. Thus, planning how to dive and look for waste is much easier in those clear waters. This does, however, sometimes make the dumping of waste occur in deeper waters, at times and in places where no one is able to see. Let me provide one example. When I took the ferry to the mainland from Koh Samui to catch a bus up north toward Bangkok, I stood beside the railing observing the waves at sunset and a while afterwards. While marveling at the incredible beauty of the surroundings, with the horizon of water turning darker and darker with the diminishing light, and small islands lighting up in the distance, I noticed a shadow on the upper deck of the ferry. The shadow moved quickly toward the railing and threw a big black bag out into the air. The bag made a circular movement down toward the water before breaking the surface with a big splash, and then it was gone. Thus, rather than keeping the bag of trash until the ferry reached land in about an hour and disposing of it properly, the solution was to throw it overboard. This bag will eventually break down, and the mass of waste materials within it will spread out along the bottom, or move away with the water, depending on the buoyancy of the material (as I discuss in Chap. 4). This partly explained to me how so much waste could be floating on the waves close to shore, even if I did not see people throw waste into the water. This dumping took place in obscure places, where no one was meant to see it.

Regarding temperature, in Sweden the water holds 2–10 degrees Celsius for most of the year, and in Thailand and Brazil 20–25 degrees Celsius. This makes for a rather different diving experience. In Sweden most divers will dive in a drysuit to keep warm, whilst in Brazil and Thailand most divers use wetsuits. When you dive using a drysuit you need to control the buoyancy of both the suit and the BCD. There are more cables to keep track of, and there is a greater risk of suit puncture if a diver gets caught in sharp waste. Hence, additional equipment adds risks to the dive, but is also what makes cold water diving possible for most parts of the year.

The difference that visibility and temperature make during a dive can be summarized by how one diving instructor in Arraial do Cabo reacted after I told him that I learned to dive in Sweden. "Great! That means that you will not panic if visibility goes down, and even if this is a cold dive for being in Brazil. It is 18 degrees Celsius, and for me this is cold!" I asked him what he meant when he said that I will not panic, and he replied,

> Well, on this dive we sometimes encounter currents that whip the sand up in the water, reducing visibility from 15 to 20 meters down to around 10 meters. For divers who have only dived in South America or Southeast Asia, 10 meters can be a scary low. But when I have Nordic divers with me, you usually tell me afterwards what nice visibility we had. That's the difference!

Different experiences of diving and water thus impact the way divers react to varying conditions below water, and this also affects how divers interact with other divers, instructors, or even with the environment. As noted earlier, diving can test a diver mentally, and if a diver cannot stay calm, accidents can happen. It also creates a greater risk that a diver may react in ways that will damage the environment, even if this is not the intention, such as kicking corals due to a panic reaction while trying to move quickly in the water.

Different Waters?

In this chapter we have engaged different bodies of water, namely, Lake Mälaren, the Baltic and North Sea, the South Atlantic Ocean, and the Gulf of Thailand. These bodies of water have different levels of visibility, hold different temperatures, and have different levels and kinds of pollution. However, are these, therefore, unrelated to another? If you pollute in one, do you not pollute the others? If you clean in one, do you not also clean in another? In all examples provided above, and in this **book** at large, the idea I try to convey is that all water is one. This does not mean that it is the same. But if you take care of water in your specific location, you do by extension also care for water elsewhere. Plastics and other waste move with water, get eaten by birds or other animals or organisms that travel, or are pulled up or processed by humans to be dumped elsewhere. Smaller pollutants can travel with rain, or storms, or as ballast water in ships. Thus, pollutants and waste move around the world. By making these substances and materials occur less in the environment, on land, in and beneath the water, then the world is cleaned one piece and particle at a time. Furthermore, another similarity among all groups with which I have worked is the idea that educating the general public and making authorities assume responsibility prevents more waste and pollutants from reaching water, and thus makes cleaning the water feasible. For, as long as waste enters the water at the current level, the battle against waste in the water cannot be won. It is by making this waste, and these practices, visible that the situation can change. Or so my interlocutors from Sweden, Brazil, and Thailand argue.

References

Ahmed, S. 2014. *The Cultural Politics of Emotion*. Second Edition. Edinburgh University Press.

Bastian, M., and T. van Dooren. 2017. Editorial Preface. The New Immortals: Immortality and Infinitude in the Anthropocene. *Environmental Philosophy* 14 (1): 1–9.

Branch, J. 2016, July 27. Who Is Polluting Rio's Bay? *New York Times*. https://www.nytimes.com/interactive/2016/07/28/magazine/rio-sarapu-guanabara-bay-pollution.html. Accessed 26 July 2023.

Cassaniti, J., and J. Hickman. 2014. New Directions in the Anthropology of Morality. *Anthropological Theory* 14 (3): 251–262.

Fassin, D., ed. 2012. *A Companion to Moral Anthropology*. Oxford: Wiley Blackwell.

Kleinman, A. 2006. *What Really Matters: Living a Moral Life Amidst Uncertainty and Danger*. Oxford, UK: Oxford University Press.

Liboiron, M. 2021. *Pollution Is Colonialism*. Durham and London: Duke University Press.

Mathias, J. 2020. Sticky Ethics: Environmental Activism and the Limits of Ethical Freedom in Kerala, India. *Anthropological Theory* 20 (3): 253–276.

Musa, G., and K. Dimmock. 2012. Scuba Diving Tourism: Introduction to the Special Issue. *Tourism in Marine Environments* 8 (1–2): 1–5. https://doi.org/10.3727/154427312X13262430523947.

———. 2013. The Future and Sustainability of SCUBA Diving Tourism. In *SCUBA Diving Tourism*, ed. G. Musa and K. Dimmock. London and New York: Routledge.

Pandian, A. 2021. Leverkusen. *Otherwise* 2, May 26. https://www.otherwisemag.com/leverkusen. Accessed 26 July 2023.

Rodineliussen, R. 2017. *Divers Engaging Policy—Practices of Making Water*. Master's thesis. Stockholm University and Sida. http://urn.kb.se/resolve?urn=urn:nbn:se:su:diva-147293.

———. 2019. Ethnographic Exploration of a Sensorial Underwater World. *Irish Journal of Anthropology* 22 (1): 184–189.

Warner, M. 2002. Publics and Counterpublics. *Public Culture* 14 (1): 49–90.

Zigon, J. 2008. *Morality: An Anthropological Perspective*. Oxford, UK: Berg.

CHAPTER 9

Ending

Sitting in my apartment in Stockholm holding a cup of warm coffee and looking out through the window, the stories I tell above seem distant. It is odd to think that the water I just used to brew this cup of black coffee would be the result of the practices and aquabiopolitical relations I have been scrutinizing over the past few years. However, the water could possibly (in fact, most likely does) contain microplastics particles and other toxicants that do not appear in the authorities' analyses, and are not caught in the cleaning procedures of the water treatment plant. These drops of water also connect me to the various bodies of water where, together with the divers from Rena Mälaren, I have crawled around along bottoms in search of toxic waste; or the areas where I together with marine scientists have sampled water and sediment in order to map levels of pollution. Urban water in society is bound up with circular relations of cleaning, use, pollution, and cleaning again—with losses and gains along the way. In what follows I attempt to highlight some of the key points made in the previous chapters, following the three main research questions I raise in the Introduction. I will end the book with the most recent developments for the trash scuba divers, showing how their current work ties even closer ties with the scientific community in their joint quest for healthy water. However, before we begin, let me reiterate what I have called aquabiopolitics and how it is applied in my analysis.

R. Rodineliussen, *Underwater Worlds*, Sustainable Development Goals Series, https://doi.org/10.1007/978-3-031-63370-6_9

Aquabiopolitics comprises the relationships between humans, waste, pollution, and water. It shows how humans govern life in water with their own well-being on land as the primary objective. Still, although humans are the driving force in these relationships, they are not the only active part. Organisms and animals react to man-made changes in these relationships, in ways not always aligned with human aims. Aquabiopolitics highlight those relations that create the underwater world of today, and suggest what might be done to act on these relationships in order to create a livable future for humans and other-than-humans alike.

How Can One Show the Human Harm Caused to the Baltic Sea and Lake Mälaren?

In an attempt to answer this question, I have accompanied marine scientists to learn from and with them about ways of engaging the world below water. The scientists I joined looked on a number of topics, ranging from the eutrophication of water, methane production in sediments, topographic changes of the seafloor resulting from climate change, and microplastic toxicity in water. These scientists utilized many technical devices in their work, and in the end, they did their best to communicate their findings to the general public.

According to these scientists, the human impact on the Baltic Sea and Lake Mälaren is devastating and leads to dead bottoms following eutrophication, escalating methane production that contributes to climate change, and overall pollution of the water. By providing fact-based evidence, they try to educate authorities and the public, and bring about more sustainable ways of treating water in the future. However, not all of their findings have been negative. For example, they have found that levels of eutrophication are decreasing in the Baltic Sea, and that it might be possible for the sea to recover if it is treated carefully from now on. Thus, they argue for an urgent societal change on a systemic level if we are to act on the opportunity to revive the Baltic Sea and save it from a certain death.

As a counterforce to these efforts, there have been a growing distrust of scientific statements about the environment among the general public. And fact-based knowledge in itself has not been enough to create public and governmental concern on matters regarding waste in water. Here the work of the trash scuba divers has created a new way to engage authorities and the public with waste in a hands-on experience, complimenting the evidence provided by the scientists.

My research with the trash scuba diving organization Rena Mälaren has shown that, in addition to sharing knowledge about what waste in water does to the marine environment, it is important that people on land get to engage with this waste because doing so evokes the emotional impetus to act. Since 2019 the divers have collected an enormous amount of waste from the bottoms around Stockholm. This waste has been displayed in piles, and made into sculptures and art to interact with people in different ways. An 8-meter-tall replica of the Stockholm City Hall made from waste was actually floated on a barge in front of the real City Hall during two summer days in 2020. Such installations confront people with the waste, causes them to think about why it has ended up in the water, what type of work has been done to remove it, and what can be done to prevent more waste from finding its way into the water. These installations manifest what I call aquabiopolitical relations when they show how humans have used water as a dumping ground, not caring about the impact of waste on aquatic life. It is aquabiopolitical relations because it is not the damage to life below water that people react the strongest to; they are appalled by the fact that all this toxic waste is within the areas from where drinking water is taken for the inhabitants of Stockholm. Thus, when the local politicians of Stockholm become involved in cleaning Lake Mälaren from waste, following demonstrations by the trash scuba divers, it can be seen as the authorities acting to change the human–waste–water relationship, with human well-being in focus. The divers are aware of this human bias, and use it to capture the attention of the authorities and the public.

Beyond creating situations where people can physically interact with waste, scuba divers and scientists utilize the opportunities for outreach provided by regular and social media. Here the role of images play a vital role. A carefully chosen image can show people things they have never seen for themselves. However, images can also be manipulated, or there may simply be too many of them, so that individuals have a hard time relating to them all. Hence, it is the combination of visual images *and* stories broadcast on social and traditional media, together with physical installations *and* interactions with waste that has made the general public in Stockholm increasingly aware of the damage done to their marine environment. Media have been important to increase outreach, but the physical installations and interactions with waste are equally powerful in making people start caring about keeping the water clean from waste.

What Is the Role of Technology in Learning About the Marine Environment?

Technology is vital to both marine scientists and scuba divers because it allows them to engage with the underwater world. Scientists use instruments to sample different layers of water and sediment at specific places and times. Other technical devices allow them to bring these samples up from the water in order to analyze them on land. Moreover, the actual analysis is further done using instruments that provide answers and calculations to what these sediments and water samples contain, such as levels of phosphorous or methane. Machines and technical devices are further used to simulate the undersea world where microorganisms such as methanogens, bacteria, and even plastics, can be analyzed as though they were still in the Baltic Sea or another body of water. To provide some examples, we can think of the methanogens Paula investigate at Radboud University, described in Chap. 2, or the way that the buoyancy of plastics particles is investigated by the scientists at SINTEF in Norway, from Chap. 4. In both of these cases, scientists utilized the capacities of instruments and technical devices to facilitate "real-world" scenarios wherein different experiments could be conducted to better understand the way these organisms or plastics react to different scenarios. This enables scientists to think about ways to counter real-world problems resulting from aquabiopolitical relations: increasing methane production from the oceans (rather than the oceans absorbing greenhouse gases), eutrophication that leads to dead bottoms, and the escalating amounts of plastics circulating in the water around the world.

Scuba divers cannot operate without the assistance of advanced machinery and other devices, as a matter of fact, scuba diving is only possible due to the scuba technology. It is the regulator that allows humans to breathe below water, the suit that keeps their bodies warm and protected from the water, the BCD that allows them to move around in the water comfortably, and not least, the goggles that allow them to look around and see the world below water with their own eyes. For this reason, I argue that scuba divers in fact can be seen and analyzed as "cyborgs"—as humans and technology working as one.

However, when thinking about technology and scuba diving, it is not only advanced machines that ought to be mentioned. There are also other forms of technologies that, at least for the group of trash scuba divers among whom I have worked, are vital for their operations. I am here thinking about the rope. The rope has become a lifeline between a scuba diver and the so-called rope puller. It is also the rope that is used to move waste objects from the water to land. Neither the scientists nor the divers can do what they do without technology. Thus, technologies should rightfully be considered as having a crucial role in the production and dissemination of knowledge, as well as in opening opportunities for humans to engage with the world below water.

In What Ways Can the Practices of Marine Scientists and Environmental Activists Impact Multispecies Care and Prevent the Destruction of Aquatic Worlds?

Thinking about aquabiopolitical relations allows one to go beyond the human-centered view of water and shift the perspective from above to below. If this is done, one can understand that in order to create a livable world for humans and other-than-humans for the future, below or above the surface, one must view the world holistically. To illustrate, let me use one example that I presented in Chap. 1 that show what happens when the perspective is too narrow and human-centered. Namely, that of how humans overuse nutrients on farms to increase crop production. These excess nutrients travel from farm to sea and raise eutrophication to harmful levels. At this stage, aquabiopolitical relations become visible. Human societies have lately addressed the issue of eutrophication, desiring to better the situation, but this is done with a focus on improving the lives of humans, rather than considering other-than-humans. Other-than-human life in water has been solely managed to improve human life on land.

In contrast to the human-centered perspective exemplified above, the scientists and environmental activists I have worked with are engaged in ways of investigating the underwater world that might alleviate the situation for the future. These scientists and activists are pursuing solutions that not only consider human well-being. In the stories they tell, other-than-humans are allowed center-stage in a number of ways. While these scientists and activists also care for humans, they believe that in order to turn things around, to work toward a future where water will be healthy and still be able to provide humans and other-than-humans means to thrive, one must also think beyond humans and start to care for the water and all the beings within it.

We have seen this philosophy in action. For example, when the trash scuba divers collect waste from the bottoms, they are careful not to remove objects that are not directly toxic if they have become a habitat for marine life. And if, by mistake, snails or other living beings are brought out of the water, the divers take pains to return them to the water. However, the divers must remove as much *toxic* waste as possible, even when it has become a habitat for marine life. Although these waste objects can provide shelter, they simultaneously expose the marine beings to harmful toxicants. Another example is the way Martijn thinks about worms and cable bacteria in the marine sediment. His work shows their importance in bringing oxygen down into the sediment which is vital for much life and many processes there. If worms and cable bacteria enjoy the right conditions, they not only will thrive but can also contribute to creating a livable future for other beings in water, and by extension, for humans on land.

Most processes that are harmful to marine life are not visible to the naked eye, or at least, not immediately. These slow processes can well be described as slow violence. In our discussion on car tire toxicity, it became clear that due to car tires degrading very slowly while under water, once dumped, they will remain in the aquatic world for a long time, continuously leaking toxicants. However, since most car tires are thrown into the water where people on land cannot see them, and the toxicants leaked are invisible, this is something most humans do not think about. When scientists and environmental activists bring this issue to fore by creating installations from car tires, sharing images of the waste on social media, and exposing the toxicity, this slow violence is made visible to those above water.

Of course, one must be realistic about how much change environmental activists can achieve. In the case of Stockholm, some changes have already materialized. As we have seen, the City has started to collaborate with the divers in collecting the waste brought up from the water. Some companies have also made changes in the way they operate as a result of discussions, and sometimes work, with the divers. One example is the electric scooter companies. Following pressure from Rena Mälaren, they have made it impossible to park their scooters close to the water (a blocking mechanism of a few meters has been installed), and this has led to fewer scooters being found in the water. Change is slow, but there is visible progress—incremental, but visible.

The environmental activists and the scientists I have worked with are constantly seeking ways of making humans care about the aquatic world and the beings within it. They hope that by confronting people with the effects of human mistreatment of the marine environment, it will be possible to create new, more sustainable, relationships between humans and water. One critical aspect of this work is to make people realize that they are part of the problem, and that it is not enough to pay for fixing the situation somewhere else; they must change the way they treat their own waste and water. The problem and the solution are not only in another place or another time, but right here, in people's everyday life. My interlocutors argue that while one person cannot fix the world, everyone can make a beginning, and they seek to lead by example.

Ending with an Example and an Update

One such example is how the divers of Rena Mälaren have started to work with seagrass and similar water plants to develop their care of water from removal of physical waste to work against eutrophication levels in Lake Mälaren and the Baltic Sea. The group has been so effective in their removal of underwater waste that most areas of inner-city Stockholm are now relatively clean from waste on the bottoms (although new objects continue to appear). In order to keep up the work, they have started to look toward other types of pollution. As one of the marine biologists among the divers told me when discussing the possibility of steering the work in this direction:

> Seagrass and similar plants are very important for the underwater world as they are effective in capturing excess nutrients from the water, thus actively working against eutrophication. Moreover, these plants are very efficient in capturing carbon dioxide from the atmosphere, as well as providing a nesting habitat for small fish. They can make a big difference if we can make them return.

These capacities of, for example, seagrass have been noted in a recent UN report, *Out of the Blue* (2020), wherein seagrass qualities are described and emphasized. The report shows how, although only occupying 0.1% of the seafloor, seagrass still stands for 10–18% of the ocean's total carbon capture, and its role in lessening the impacts of eutrophication is emphasized. Moreover, the report shows that even though seagrass provides such important ecosystem services, human activities are rapidly killing off the plants from the ocean floor. It has lost more than 30% of its habitat over the last decade. The marine biologist diver further explained,

> Replanting seagrass and similar plants is not easy. And the plants often grow slowly, it can take up to 50 years before they reach their full potential. And still, planting such plants is important because it can have beneficial impacts in a longer perspective. But mainly, because doing so while using our media coverage can pull attention to the importance of helping seagrass and similar water plants across the globe. This would have an immediate impact on climate change, among other things. By making people talk about water plants, we can help these plants to survive, and thus continue providing all their important services to life below and above water.

The scuba divers seek to both nurture existing water plant meadows and replant them in areas where they grew earlier. They look toward similar attempts along the Swedish west coast for inspiration. Their idea is to plant in areas they have cleaned from waste. Then they will monitor and document how the plants grow (or not), to see if this can be done effectively in an urban underwater environment. Meanwhile, they will seek to make good use of their social media following to add their voice to those advocating for the importance of caring for water plants globally. With this work, the scuba divers make aquabiopolitical relations visible. They highlight how humans continuously keep damaging the underwater world at the expense of human pleasure. Seagrass meadows are, for example, often located in good spots for boat marinas and are therefore removed. The divers seek to change such relations between humans and the marine environment, hoping that if they show how we can care for water plants—and how important it is to do so—it might inspire others to also think outside their own bubble, help people make different choices, which could lead to other sorts of relations between humans and the underwater world.

Furthermore, such outreach work aligns with the aim of Rena Mälaren to expand beyond Stockholm, and ultimately, outside of Sweden. The group has come to realize that in order to keep increasing their environmental impact they need to look beyond the capital and engage in international work for the environment. I provide some examples from such initial expansion work in Chap. 8, where Rena Mälaren traveled to other cities in Sweden. But what I do not tell in that chapter is that Rena Mälaren has decided to change their name to *Hands2Ocean*. The change came into effect in early 2024. This decision comes with the explicit aim of expanding nationally and internationally. Fredrik has, somewhat ironically, motivated the change of name by saying,

> We can no longer have a name only focusing on Lake Mälaren. For example, when we go to Gothenburg, they always laugh at us saying that our compass is not right and that we are not even close to Mälaren. We cannot have it like this, haha! We need a new name and one that works on a global stage. Hands2Ocean works internationally, and it is a name that speaks to our work, as we lend our hands to helping the ocean.

Thus, from here on Hands2Ocean will work to expand cleanup operations to other cities in Sweden, and ultimately to other countries. Meanwhile, they will further engage in work with water plants to address the problematic situation with eutrophication, in the Baltic Sea and in Lake Mälaren. While they do this, they will also be part of a larger project of making the world care about water plants, like seagrass, as they are of great importance for the well-being of our oceans, and in the long run, that of humans. This new direction of the group will inevitably lead them toward closer collaboration with the marine sciences, as this is where the scuba divers find guidance on *how* to address marine pollution like eutrophication.

To Sum Up

This **book** includes stories from 2017 to 2024. They are focused on the water around Stockholm: Lake Mälaren and the Baltic Sea. But there are also narratives from the Netherlands, Norway, Brazil, and Thailand. These accounts show that although waste management practices might differ around the world, and the amounts and types of waste in water varies considerably, there is one great commonality: water has always been, and still is, seen as a dump for unwanted things. If we look at the relationship between societies, waste, pollution, and water from an aquabiopolitical perspective, such processes and relationships can be made visible, and that is the condition for the possibility of changing them toward relationships of care rather than negligence. To give the last word to the divers of Hands2Ocean: *Water is Life. Make it Important!*

Reference

United Nations Environment Programme. 2020. *Out of the Blue: The Value of Seagrasses to the Environment and to People*. Nairobi: UNEP.

Appendix A: Inventory—What Can Be Found on the Bottoms Around Stockholm

Total count between mid-2019 and early 2024: **348,000 kilograms**

Most common items:
Car batteries (by early 2024 39,000 kilograms collected)
Car tires (the Damn Goat alone consisted of 700 tires)
Cigarette butts
Bicycles
Electric scooters (by early 2024 1262 units collected)
Plastic bags
Plastic bottles
Diapers and wet wipes
Aluminum cans
Glass bottles
Scrap metal
Pipes
Road signs
Telephones and smartphones
Fishing tackle

R. Rodineliussen, *Underwater Worlds*, Sustainable Development Goals Series, https://doi.org/10.1007/978-3-031-63370-6

Home appliances (refrigerators, freezers, toasters, microwaves, radios, etc.)
Scooters
Wallets
Clothing
Power tools
Commodes

Less common but reoccurring items:
Cars (by early 2024 8.5 cars had been brought out of the water with the help of tow trucks)
Weapons (explosives, handguns, knives)
Drugs (mainly marijuana)
Ship wrecks (old and new)
Sex toys

Index[1]

[1] Note: Page numbers followed by 'n' refer to notes.

R. Rodineliussen, *Underwater Worlds*, Sustainable Development Goals Series, https://doi.org/10.1007/978-3-031-63370-6

W

The manufacturer's authorised representative in the EU is Springer Nature Customer Service Centre GmbH, Europaplatz 3, 69115 Heidelberg, Germany. If you have any concerns regarding our products, please contact ProductSafety@springernature.com

Printed and bound by CPI Group (UK) Ltd, Croydon, CR0 4YY
01/10/2026
02225904-0001